T0305501

Understanding Data Analytics and Predictive Modelling in the Oil and Gas Industry

This book covers aspects of data science and predictive analytics used in the oil and gas industry by looking into the challenges of data processing and data modelling unique to this industry. It includes upstream management, intelligent/digital wells, value chain integration, crude basket forecasting, and so forth. It further discusses theoretical, methodological, well-established, and validated empirical work dealing with various related topics. Special focus has been given to experimental topics with various case studies.

Features:

- Provides an understanding of the basics of IT technologies applied in the oil and gas sector.
- Includes deep comparison between different artificial intelligence techniques.
- Analyzes different simulators in the oil and gas sector as well as discussion of AI applications.
- Focuses on in-depth experimental and applied topics.
- Details different case studies for upstream and downstream.

This book is aimed at professionals and graduate students in petroleum engineering, upstream industry, data analytics, and digital transformation process in oil and gas.

Understanding Data Analytics and Predictive Modelling in the Oil and Gas Industry

Edited by
Kingshuk Srivastava, Thipendra P Singh,
Manas Ranjan Pradhan and Vinit Kumar Gunjan

CRC Press is an imprint of the
Taylor & Francis Group, an **informa** business

First edition published 2024
by CRC Press
6000 Broken Sound Parkway NW, Suite 300, Boca Raton, FL 33487-2742

and by CRC Press
4 Park Square, Milton Park, Abingdon, Oxon, OX14 4RN

CRC Press is an imprint of Taylor & Francis Group, LLC

ISBN: 9781032413877 (hbk)
ISBN: 9781032413891 (pbk)
ISBN: 9781003357872 (ebk)

DOI: 10.1201/9781003357872

Typeset in Times
by Deanta Global Publishing Services, Chennai, India

In loving memory of our dear colleague and one of the contributors of this book, Dr. Bhagwant Singh

Contents

Preface

The hydrocarbon industry has always been one of the most prominent investors in the development and funding of new and cutting-edge technologies after the defence and medicine sectors. From exploration, production, and refining to even transportation of crude implements a lot of high-end technology is implemented. The most widely used technology in every aspect of the hydrocarbon value chain is the combination of electronics and information technologies.

The application of different IT technologies is highly pervasive in the different processes in the field of oil and gas which has enabled the concepts of automation, Industry 4.0, or digital oilfields. In the last couple of decades some of the most cutting-edge IT technologies have made life a lot easier in all the aspects of an individual with some radical and pathbreaking outcomes. Most of it is possible due to universally compatible and advanced hardware made available in the market at a very cheap rate.

One of the most important parts of any business's success is the competitive edge it has over its rivals and the ability to correctly predict future market scenarios. This enables the company to have foresight in the market and be able to plan accordingly. Now, with the advent of technologies like cloud computing, big data, data mining, machine learning, and statistical methods, concepts such as predictive/prescriptive modelling have been able to give highly accurate forecasts of different aspects of market dynamics. This further enables an organization to cut costs, redefine their strategy, and increase profit.

Both oil and gas business and computer science-based analytical modelling are very different fields of expertise. They need years of practice and in-depth study to be able to implement the logics and algorithms for any worthwhile outcome. But any program or system deployed in a particular business domain needs to understand the basic mechanism and working principles of that field of application, without which the computer programs could not be properly designed and practically implemented. There has been high demand of professionals/experts who have knowledge about both the domains with practical experience in these two extremely evolving and high-cost domains. Although there are separate books dealing with each domain, catering to the needs of this combined field, a book which can combine the different attributes of both fields and describe the exact application-level knowledge is almost non-existent. This has been the major motivation behind conceptualizing and designing this book.

The biggest challenge which presented itself while developing this book was to get contributors for the different chapters. The problem was same as stated above: there are very few experts available in the market today who can contribute to this niche field of technological implementation. Again, these experts are already overloaded with too much of work, so convincing these people to find time and contribute to a book was one of the biggest challenges faced by the editors.

In our professional career of IT, it has always aggravated us that there are hardly any good quality books which can be referred to when stuck at any problem during IT applications in the field of oil and gas. This has been the inspiration behind designing, contributing, and editing this book.

The book is based on the practical approach of different analytical processes which are further explained with multiple case studies in each chapter, wherever it is required. This enables the reader to easily internalize the application of different algorithms of analytics in the field of oil and gas. The whole value chain of the hydrocarbon industry is described: upstream, midstream, and downstream with different technologies applied at the operational level.

The main aim of this endeavour was to design an amalgamation of two of the most highly evolving and volatile technologies and present it in a form which could be easily understood, and the knowledge thus acquired could be applied in practicality.

We sincerely hope that this effort could bring the desired results.

Kingshuk Srivastava
Thipendra P Singh
Manas Ranjan Pradhan
Vinit Kumar Gunjan

About the Editors

Kingshuk Srivastava is working as a faculty at UPES, Dehradun, India with over 12 years of extensive experience of data analytics in the oil and gas sector. He has earned his Ph.D in computer science engineering from UPES and his field of research is in AI, data science, and NOSQL databases. He has designed, developed, and delivered training to Birlasoft Technologies on different IT aspects in the oil and gas value chain. He has published multiple papers in his field of interest in national and international journals.

Thipendra P Singh is currently positioned as a Professor in the School of Computer Science Engineering and Technology, Bennett University, Greater Noida, NCR, India. Prior to this, he has been associated with UPES University and Sharda University. He holds Doctorate in Computer Science from Jamia Millia Islamia University, New Delhi. He carries 26 years of rich experience with him. He is supervising PhD scholars with French and UK universities also. He has been a widely traveled academician and participated at various platforms across the countries including UK, France, UAE, and Singapore. He has been the editor of 10 books on various allied topics of Computer Science. Dr. Singh is a senior member of IEEE and a member of various other professional bodies including IEI, ACM, EAI, ISTE, IAENG etc., and also on the editorial/reviewer panel of different journals. He is a fellow of IETA-India since 2019. He is also on the board of studies of different Indian and abroad Universities.

Manas Ranjan Pradhan holds a Ph.D (Computer Science) from University of Mysore, India, and Master of Technology (Computer Science) from Utkal university, India. He has vast experience in teaching, research, and academic administration in India and abroad. He is currently working at Skyline University College, Sharjah, UAE. He has been associatied with the IT industry for industry-academic collaboration, internship, placement, and workshops. He has executed the IBM Center of Education for Cloud Computing and Business Analytics at INTI International University, Malaysia under Laureate International Universities, USA. He has presented and published many research papers in various conferences and journals. He has three Indian patents and three Australian patents to his credit. His areas of expertise are business analytics, datamining, data warehouse, retail/ecommerce analytics, artificial intelligence, machine learning, and business process modelling.

Vinit Kumar Gunjan is Associate Professor in the Department of Computer Science and Engineering and Dean of Academic affairs at CMR Institute of Technology Hyderabad. He has published research papers in IEEE, Elsevier, and Springer Conferences, authored several books, and edited volumes of the Springer series, most of which are indexed in the SCOPUS database.

Contributors

Utkkarsh Agarwal—Department of Petroleum Engineering and Earth Sciences, University of Petroleum and Energy Studies, Dehradun, India

Tanupriya Choudhury—Professor, CSE Department., Symbiosis Institute of Technology, Symbiosis International University, Lavale Campus, Pune, India

Utkarsh Das—Department of Petroleum Engineering and Earth Sciences, University of Petroleum and Energy Studies, Dehradun, India

Shaurya Gupta—School of Computer Science, University of Petroleum and Energy Studies, Dehradun, India

K.T. Igulu—Department of Computer Science, Kenule Beeson Saro-Wiwa Polytechnic, Bori, Rivers State, Nigeria

Madhu Khurana—University of Gloucestershire, UK

Devarani Devi Ningombam—Department of Informatics, SoCS, University of Petroleum and Energy Studies, Dehradun, India

Atul Kumar Patidar—Department of Petroleum Engineering and Earth Sciences, University of Petroleum and Energy Studies, Dehradun, India

Z.P. Piah—Department of Computer Science, Kenule Beeson Saro-Wiwa Polytechnic, Bori, Rivers State, Nigeria

Monideepa Roy—School of Computer Engineering, KIIT Deemed to be University, Bhubaneshwar, India

Achala Shakya—School of Computer Science, University of Petroleum and Energy Studies, Dehradun, India

Vinod Kumar Shukla—Department of Engineering and Architecture, Amity University, Dubai, UAE

Bhagwant Singh—Department of Informatics, SoCS, University of Petroleum and Energy Studies, Dehradun, India

Priyanka Singh—Department of Informatics, SoCS, University of Petroleum and Energy Studies, Dehradun, India

Thipendra P Singh—School of Computer Science Engineering and Technology, Bennett University, Greater Noida, India

Kingshuk Srivastava—Department of Informatics, SoCS, University of Petroleum and Energy Studies, Dehradun, India

Venkata Sravan Telu—Applied Computer Science Graduate School, Northwest Missouri State University, Maryville, USA

Gaurav Tripathi—Department of Geoinformatics, Central University of Jharkhand, Ranchi, India

Sonali Vyas—School of Computer Science, University of Petroleum, and Energy Studies, Dehradun, India

Noore Zahra—Department of Computer Science, College of Computer and Information Sciences, Princess Nourah bint Abdulrahman University, Riyadh, Saudi Arabia

Acknowledgments

Editing a book is harder than I thought. None of this would have been possible without so many persons behind me. My mum, dad, and wife have always been the pillar of my strength and the motivation and support behind my every endeavor. This book would not have been possible without my senior colleagues and my friends from UPES, Dehradun, India.

Dr. Kingshuk Srivastava

The editor wants to thank his family, colleagues, and all well-wishers who helped him in achieving this milestone. He also wants to put on record his appreciation for his previous organization, University of Petroleum & Energy Studies (UPES) and his current organization Benett University and their managements to provide such a wonderful opportunity to let him grow academically. Above and over, he feels thankful to the Almighty for showering all His blessings upon him.

Dr. Thipendra P Singh

The editor is thankful to all the authors for their immense contribution to the completion of this book. He also thanks his other editor friends, colleagues, and family members for their continuous support. He acknowledges the continuous encouragement from his organization Skyline University College, Sharjah, UAE, at every stage of the book's development, and acknowledges the grace of God for the successful finishing of the book. He wishes that it will fulfil the readers' needs to a maximum extent.

Dr. Manas Ranjan Pradhan

The energy sector is constantly changing, and energy suppliers need to develop newer and newer technologies to generate, store, and transport energy to households and businesses. The energy industry is the basis of our civilization and economy. I am thankful to the publishers and creditors for joining in the creation of this timely title which will be revolutionary for human life.

Dr. Vinit Gunjan

Acknowledgments

1 Understanding the Oil and Gas Sector and Its Processes

Upstream, Downstream

Atul Kumar Patidar, Utkkarsh Agarwal, Utkarsh Das, and Tanupriya Choudhury

1.1 INTRODUCTION

Some of the most valuable commodities in the world are crude oil and gas. Their diverse spectrum includes producing fuels for transportation, energy generation, and petrochemical goods such as plastics, solvents, etc. As a result, the oil and gas sector is one of the most influential components of the global economy, and fluctuations in its price have a significant impact on most manufacturing industries as well as consumers. Standard Oil, Royal Dutch Shell, and British Petroleum are the leading supermajors of the oil and gas industry. John D. Rockefeller began his career in refining and formed the Standard Oil Company, which by the end of 1879 controlled 90% of total America's refining capacity. Today, ExxonMobil is the successor to Standard Oil. The Royal Dutch Petroleum started in the East Indies in the late 1800s, and by 1892, integrated production, pipelining, and refining operations were started. In 1907, Royal Dutch and Shell together formed the Royal Dutch Shell Group. In the same year, a British former gold miner discovered oil in Iran, which led to the incorporation of the Anglo-Persian Oil Company. In 1914, the British government purchased 51% of the company to provide sufficient oil for the Royal Navy. Later, the company became British Petroleum (BP). A cartel of seven companies was formed that controlled the world's oil and gas business. Known as the "Seven Sisters", they included Exxon, Royal Dutch/Shell, BP, Mobil, Texaco, Gulf, and Chevron. The Organization of the Petroleum Exporting Countries (OPEC) was established in the year 1960 by the governments of Venezuela, Saudi Arabia, Kuwait, Iran, and Iraq with the purpose of negotiating with Integrated Oil Companies (IOCs) about issues relevant to oil production and potential concession rights. In the 1990s, Saudi Arabia surpassed the United States as the world's largest producer. Because of the development of techniques for extracting hydrocarbons profitably from shale, a rock with a relatively low permeability, the United States climbed to become the

DOI: 10.1201/9781003357872-1

highest-producing country in the 2010s. In order to supply energy for irrigation, transportation, agrochemicals, etc., the oil and gas sector was one of the most crucial factors in the Green Revolution. As the world is moving towards a digital future, companies are embracing innovative technology to change their operational environments for increased productivity, greater efficiency, and cost savings.

The oil and gas industry can be divided into three sectors: upstream, midstream, and downstream. The upstream sector consists of various activities such as acquiring land rights from the government to conduct geological and geophysical surveys to find the hydrocarbon reservoir, drilling, and extracting the oil and gas from the reservoir. The downstream sector includes processes such as crude oil refining, processing of natural gas, and distribution of the products derived from the crude oil and natural gas. In midstream operations, the hydrocarbons are transported from upstream production to downstream refining and processing plants. In this chapter, we shall be discussing the identification of the geological origins of petroleum reservoirs and reservoir fluids, the history of the oil and gas industry, the structure of the modern oil and gas industry, various disciplines that make up petroleum engineering, the differences between conventional and unconventional reservoirs, analyzing rudimentary engineering methods that are used in exploration and production, and the interpretation of semi-log and cross-plots.

1.2 IDENTIFICATION OF THE GEOLOGICAL ORIGINS OF PETROLEUM RESERVOIRS AND RESERVOIR FLUIDS

Sedimentary rocks typically consist of petroleum reservoirs. Petroleum reserves are rarely found in fractured igneous or metamorphic rocks. High-pressure, high-temperature environments are where igneous and metamorphic rocks are formed, which are not ideal for the development of petroleum reservoirs. They lack the permeability or linked pore space required to create a channel for petroleum to flow into a wellbore. Metamorphic rocks have developed as parts of sandstone; they have undergone high-temperature and high-pressure conditions. Any hydrocarbons present in the occupied pores are cooked away due to the high-temperature and pressure conditions. A hydrocarbon reservoir cannot form without the presence of a number of necessary elements. First, there should be a presence of a source rock for the hydrocarbon. At their most fundamental level, sedimentary rocks were generated from ancient plant elements that were stored and later turned into fossil fuels. The remnants gather in a sedimentary site, such as shale, which turns into a source rock. The source rock's pressure and temperature should also be suitable for producing oil or gas from the organic mixture. Oil or gas production would not be at its best if the conditions were not ideal. Table 1.1 shows the key elements of the petroleum system.

The occurrence of a reservoir rock and a channel connecting from the source rock to the reservoir rock is the third factor. A reservoir rock can confine a volume of fluids and can support economically viable flow rates. Oil and gas are majorly formed in rocks which are difficult to access via the contemporary production methods. The flow rate of the hydrocarbon must be maintained so that the well is economically viable. The economic viability of the reservoir depends on two characteristics:

TABLE 1.1
Key Elements of Petroleum System

PETROLEUM SYSTEM ELEMENTS	DESCRIPTION
Source	A rock which is capable of generating movable quantities of hydrocarbons
Migration	After generation and expulsion from thermally matured source rocks, hydrocarbons migrate through carrier beds, faults and fractures to reservoir rocks
Reservoir	Rocks which store the hydrocarbons inside the pores
Trap	Geometric arrangement of reservoir and seal rocks in which hydrocarbons are accumulated
Seal	Rocks which hamper the escape of hydrocarbons from reservoir rocks

porosity and permeability. The measure of the storage capacity of a rock to hold fluids is known as porosity. In units of quantities, the porosity is the proportion of pore volume to total volume (bulk volume). The capacity and capability of the formation to allow the fluid to flow in a porous medium is known as permeability. The movement of the hydrocarbon needs to be restricted or else the fluid will move towards the surface due to migration, buoyancy, and other forces. There are two factors that restrict the fluid. An impenetrable cap rock comes first, followed by a trapping mechanism that stops lateral fluid movement. Deposition in sedimentary rocks leads to the development of traps. When the sediment that creates the reservoir rock is deposited in a discontinuity, the seals are created adjacent to and on the reservoir's top. Figure 1.1 depicts all the important petroleum system elements and their subsurface arrangement.

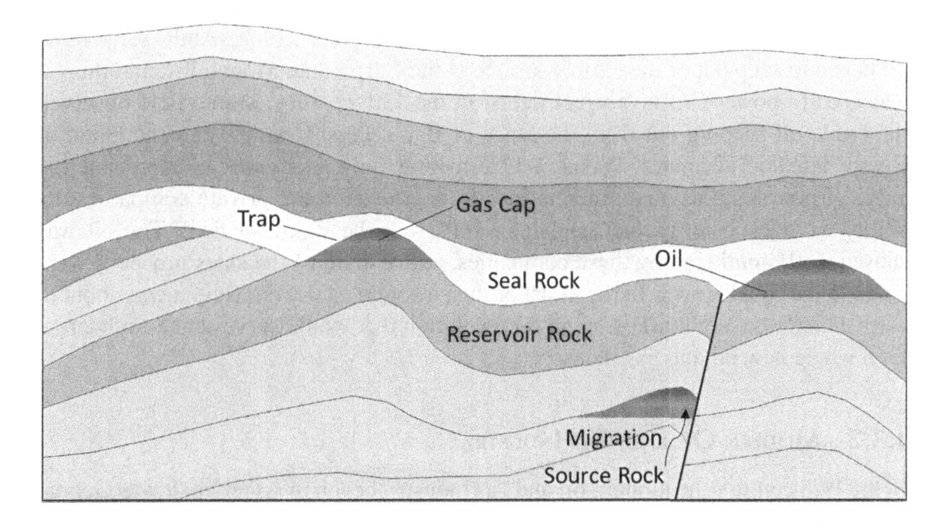

FIGURE 1.1 Conceptual diagram (not to scale) showing key elements of the petroleum system include source rock, reservoir rock, migration pathway, trap and seal.

1.3 HISTORY OF THE OIL AND GAS INDUSTRY

The history of the oil and gas industry can be divided into two parts—the ancient oil and gas industry, and the modern oil and gas industry. We will discuss each of them one by one.

1.3.1 Ancient Oil and Gas Industry

In ancient times, crude oil was used by humankind for various purposes; for example, as a water repellant, or as a binding material because of its sticking nature. Around 5,000 years ago, Sumerians, Mesopotamians, and Egyptians used natural asphalt for the construction of walls and towers, and bitumen was used to construct water canals. They also used crude oil as a binding material to join together pieces of boats for easy water transportation. The Egyptians used it to preserve dead bodies and keep them intact. By 1500 BC, humankind had learned how to use crude oil for lighting oil lamps (Hassan, 2013). Around 600 BC, Chinese people were digging salt wells when petroleum was first discovered and extracted. The Chinese historical records state that along the Tibetan border, oil wells with a depth of about 100 feet were being drilled. These wells produced both water and natural gas (Gao, 1998). Around 450 BC, Herodotus found oil pits that were located near Babylon, and by 325 BC, Alexander the Great used the flames generated from petroleum to scare war opponents, and the Roman warriors used the blazing objects that contained crude petroleum as weapons in war.

By AD 347, the Chinese had considerably developed the oil and gas industry by drilling their first oil wells up to a depth of 800 feet using bits that were attached to bamboo poles (Dalvi, 2015). By AD 1500, the Chinese started to drill the wells up to a depth of 2,000 feet and invented the pipeline transportation system to pump out the hydrocarbons from the wells to the earth's surface (Herbert, 1942). At the same time, oil began to seep out of the earth's surface at the Carpathian Mountains, and this oil was used to power Crosno's street lights. In the 13th century, Marco Polo observed the crude oil seeping out from the lands of Baku, Persia. After Peter the Great of Russia defeated Khanat of Persia in 1723, people only later came to appreciate the value of petroleum as a commercial good. Peter offered the private companies the ability to investigate, extract, and refine oil from the Baku oil field. The oil was subsequently marketed by these companies, and a share of the sales proceeds were given to the royal crown. In the midst of this, a country far from Russia was about to enter a historical era marked by an oil revolution that would serve as the foundation for a whole new oil and gas industry.

1.3.2 Modern Oil and Gas Industry

In the 19th century the modern oil and gas industry began in America. It was George Henry Bissell who laid the seed of the modern oil and gas industry globally. George Henry Bissell observed how Pennsylvanian merchants extracted oil from dams using oil-soaked blankets while searching for an alternative to asphalt-based kerosene.

Then he gathered a group of investors and recruited Professor Silliman and some Yale chemists to conduct an analysis of the viability of using rock oil as an illuminant. After accepting the project, Silliman started his investigations and the results turned out to be extraordinary. He explained that after being heated to various boiling points, rock oil (petroleum) can be distilled into a number of other fractions made of carbon and hydrogen, one of which was a very qualitative illuminant. As a result of these findings, Bissell established the Pennsylvania Rock Oil Company, the country's first oil and gas company, and this marked the beginning of the modern oil and gas industry.

Although Silliman's report was impressive, people felt that drilling an oil well was risky. But in order to start the project, the company hired the services of Edwin L. Drake, who completed the first drilled oil well at Oil Creek near Titusville, Pennsylvania, on the 27th of August, 1859. Following Drake's discovery that oil could be drilled and pumped from the soil to the surface, about five were drilled in the Creek Valley in North Pennsylvania. By 1860, around 15 refineries were built that resulted in the increase of production from 450,000 barrels in 1860 to 3,000,000 barrels in 1862. The kerosene produced from the refineries replaced other products such as whale oil and coal gas in the market. The highly qualitative illuminant which was mentioned by Silliman in his report was this kerosene. American refineries generated an estimated three times as much kerosene as was required by American markets between 1865 and 1870. This resulted in the drastic fall in the average price of kerosene. There was a continuous attempt to produce kerosene in the amount that was actually needed in the market but this could not be achieved. Hence there was a drastic fall of industry, which caused the continuous loss of money of both the oil producers and refiners. Taking advantage of the situation, Rockefeller, his partner Henry Flager, and five investors seized control of the industry by transforming their partnership into a joint stock company, the Standard Oil Company which seized 10% of the American oil and gas industry. In 1871, a year after the creation of the Standard Oil Company, the American oil and gas industry deteriorated further and the profit margins almost disappeared. During this time, Rockefeller came up with a plan to make the oil and gas business safe and profitable. His plan was to seize control of the leading refineries and take control over the smaller ones. By the time the war ended in 1879, the Standard Oil Company had control over 90% of the US oil refining industry. Between 1888 and 1891, the company began oil exploration and production, capturing over one-third of the US oil output. By the middle of the 1880s, the company had taken over 80% of the US oil market.

1.3.3 The Role of the Russian Oil and Gas Industry

Before 1870, the Czarist government's inefficiency and corruption, as well as Russia's technological and geographic isolation, severely hampered the development of its oil and gas industry. As a result, Russia's oil and gas sector did not experience its first major expansion until after 1870. The Russian government allowed for competition in business. Significant oil wells began to emerge in the Russian oil and gas sector around 1871–1872 and by 1873, more than 200 wells were producing crude oil. It is possible to argue that the Russian Nobel family contributed to the success of the country's oil and gas business. Ludwid Nobel, an industrialist, went into the oil business using the investments made by his brother, Robert Nobel. The Nobel oil and gas company at that time was producing half of the kerosene that was being used in Russia. This forced American kerosene off the Russian market, but the Russian kerosene was confined only to the local market. However, the development of a rail connection from Baku across the Caucasus to Batum, a port on the Black Sea that Russia acquired control of after its war with Turkey, was ensured by the covert entry of the wealthy Jewish financial family, the Rothschilds, into the Russian oil and gas business. Due to this, Russian oil and gas businesses gained access to the global oil and gas market, and the two Russian oil companies thereafter emerged as Standard Oil's main global rivals. The administrators of the Standard Oil Trust saw how almost impossible it would be to assimilate the Rothschild or Nobel families and they thought they could effectively compete with the Russian enterprises in Asia through the expanding oil corporation.

1.3.4 Royal Dutch Shell in the East Indies

The founder of the Dutch-based enterprise Royal Dutch was Aeilko Jans Zijlker. Even though Zijlker successfully completed his first well drilling operation on the marshy coast of Sumatra in 1885, it is reported that "the Royal Dutch Company was not launched until 1890, and the first float of its stocks was exceeded four and a half times". Unfortunately, Zijlker passed away the year the business was founded, leaving Baptiste August Kessler in charge. The company was in disarray and losing money when Kessler took over, but he turned it around in just two years, and it began turning a profit with oil output that was six times more than it had been two years previously. Due to Kessler's tenacity, commitment, and management abilities, Royal Dutch was able to develop and expand to the point that, between 1895 and 1897, the company's manufacturing capacity and sales quadrupled by five. Standard Oil, a massive American company, made numerous unsuccessful attempts to seize control of Royal Dutch. Consequently, Royal Dutch considerably and independently grew its business to become one of the top global oil firms.

1.4 THE STRUCTURE OF THE MODERN OIL AND GAS INDUSTRY

The upstream, midstream, and downstream sectors make up the petroleum industry as a whole. The upstream sector incorporates the exploration and production processes which include doing geological surveys at sites and drilling activities, both onshore and offshore, as part of production. The midstream sector is the connection between the upstream and downstream sector and between upstream production operations and downstream refining and processing, the movement and storage of oil and gas are included in the midstream sector. A substitute classification of the three-sector system is the two-sector system in which the downstream sector includes the midstream. The downstream sector involves the processing and refining of oil and natural gas and also includes the distribution of products (Fanchi and Christiansen, 2017).

1.4.1 Upstream

The upstream industry consists of activities aimed at finding, discovering, and producing oil and gas. For managing fluid production and preparing it for transportation, upstream facilities are required. This sector can be more aptly named as the exploration and production sector. The exploration stage is where the search of hydrocarbons in the subsurface happens. Land surveys are carried out to approximately pinpoint the region which holds the most potential. The geological and geophysical analyses of the subsurface are done by various techniques such as polarization surveys, drilling and logging, and the use of seismic waves. The main objective of the exploration stage is to estimate the potential and recovery factor of the reservoir. If a location has the capability to produce, exploratory wells are drilled to test the resource. When the exploratory wells are successful and the well is given the nod of being economically viable, the next step is to build production wells and start main production.

1.4.2 Midstream

The transportation of oil and gas from upstream production to downstream refining and treatment operations comes under the domain of the midstream sector. Oil and gas are transported via a variety of transportation methods. By pipeline, tanker, truck, and rail, oil and gas are transported to processing facilities and then to the end consumers. The safety of the hydrocarbon, distance travelled, and fluid state are the main factors that determine the transportation method used. For transportation of hydrocarbons pipelines have been the majorly used choice, but the construction and networking of those pipelines is a major task. Pipelines are laid for both onshore and offshore activities; These networks are normally a few thousand miles long and requires strategic placement. Long-distance fluid pipelines typically require compressor stations or pump stations to maintain fluid flow. By raising the density of the gas moving through the pipeline, gas compression helps maximize the use of available pipeline space. Trucks, trains, barges, and tankers are additional methods of transportation for oil and gas. Trucking is the most adaptable mode of land transportation because it can reach almost any location on land. Large volumes of oil can be transported by rail in an economical and effective manner. Tankers and barges transport gas and oil over waterways, including rivers and oceans.

1.4.3 Downstream

The downstream sector includes the refining, processing, distillation, and purification of oil and gas before turning it into usable and profitable products such as gasoline, fuel oils, and petroleum-based products. Starting at the well site, the generated well flow is divided into oil, water, and gas phases; the initial hydrocarbon processing takes place. Three major processes are used in refining to convert crude oil into finished products, namely separation, conversion, and purification. The initial stage of refinement is separation, and the employment of distillation towers is absolutely

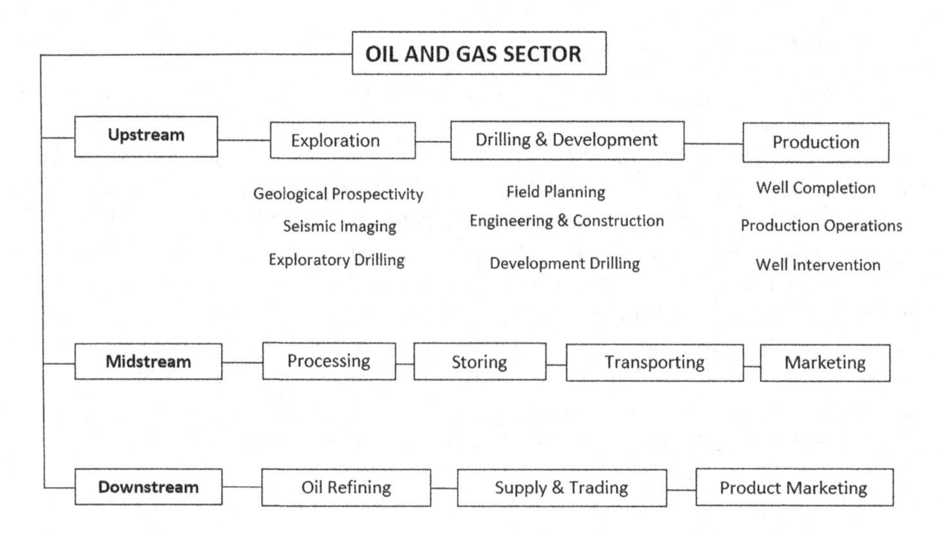

FIGURE 1.2 The structure of modern oil and gas industry is shown. It is broadly divided into three stages- upstream, midstream, and downstream which can be further divided based on the products and industries.

essential. Distillation towers, also known as atmospheric crude fractionators, based on their respective boiling temperatures, divide crude oil into its component mixes. Separation of crude oil into product streams which needs to be changed into a useful product is the separation process. In the conversion process the low-value, high molecular weight hydrocarbon mixture is converted to a high-value, low molecular weight hydrocarbon product. Conversion is accomplished by breaking higher molecular weight hydrocarbon chains to low molecular weight hydrocarbon molecules. The conversion units at which the high molecular weight hydrocarbon chains are broken are the fluidized catalytic cracker (FCC), the hydrocracker. A detailed flowchart for the different structures of the modern oil and gas industry are illustrated in the following figure (Figure 1.2). The upstream sector can be divided into the exploration, drilling, development, and production phases. The midstream sector can be divided into the processing, storing, transporting, and marketing phases. The downstream sector can be divided into the oil refining, supply and trading, and product marketing phases.

1.5 DIFFERENCES BETWEEN THE CONVENTIONAL AND UNCONVENTIONAL RESERVOIRS

A petroleum reservoir or an oil and gas reservoir can be defined as a porous and permeable formation which contains individual banks of hydrocarbons confined by impermeable rock or water barriers and is characterized by a single natural pressure system. A petroleum reservoir is formed when kerogen (ancient plant matter) in rapped rocks of the formation is converted into hydrocarbons when the rocks are subjected to high pressure and temperature. Petroleum reservoirs can broadly be classified as conventional and unconventional reservoirs.

1.5.1 Conventional Reservoirs

In conventional reservoirs, naturally occurring hydrocarbons such as oil and gas can be produced at economically viable flow rates, or we can say that economic volumes of oil and gas are produced from conventional reservoirs without using any stimulating treatments or specialized recovery methods (Fanchi and Christiansen, 2017, p. 29). In these reservoirs, the hydrocarbons are typically accumulated in favourable structural or stratigraphic traps in which the formation is porous and permeable, but they are sealed by overlying rock formations that have lower permeability, which prevents hydrocarbons from escaping. Since the formations have exceptional reservoir quality and suitable subsurface migration paths that connect the source rocks to the reservoirs, hydrocarbon production typically does not require extensive stimulation. Sandstone or limestone make up most of the sedimentary rocks found in conventional oil and gas reservoirs. Both have a wide range of lithologic differences, including original porosity, secondary porosity, or a combination of the two, and can be fine, coarse, thinly laminated, or massively bedded.

1.5.2 Unconventional Reservoirs

Unconventional reservoirs refer to the formations from which hydrocarbons cannot be produced at economic rates or do not produce economic volumes of oil and gas without stimulation treatments or special recovery processes and technologies. These reservoirs hold vast amounts of oil and gas, but their technological production presents a technical challenge to engineers and geoscientists equally (Fanchi and Christiansen, 2017, p. 29). The special recovery processes include horizontal drilling and hydraulic fracturing. In unconventional reservoirs, the rocks have high porosity and low permeability, which keeps the hydrocarbon trapped in one place; they do not require a cap rock, so the production is done directly from the source rock itself. These reservoirs are widely spread across various stratigraphic layers and basin types, storing significant amounts, and providing long-term potential for energy supply. A brief comparison of elements and related processes is shown in **Table 1.2**.

The surge in the worldwide demand for production of oil and gas from unconventional reservoirs has generated considerable interest in many countries, one of which is Saudi Arabia. Various studies have indicated that the supply and demand balance can be achieved by exploiting unconventional resources located in the major hydrocarbon basins of the world. North Arabia is producing unconventional gas from a gas field located onshore in Saudi Arabia and is operated by Saudi Arabian Oil (Sahin, 2013). The Ghawar Oil Field is by far the largest unconventional oil field in the world, and it accounts for almost half of the cumulative oil production of Saudi Arabia. A petroleum resource triangle is shown in Figure 1.3, which illustrates the history of hydrocarbon exploration through conventional and unconventional techniques.

TABLE 1.2
Contrasting Features Between Unconventional and Conventional Reservoirs

ELEMENT/PROCESS	UNCONVENTIONAL RESERVOIR	CONVENTIONAL RESERVOIR
Proximity to mature source rocks	Close	Close or distant
Migration of petroleum	Gas trapped in place	Migration may occur over a long distance
Reservoir trap	No evidence of trap	Structural, stratigraphic or combination trap
Hydrocarbon charge area	Prevalent through a large area	Relatively limited charge area
Resource in place	Large	Relatively small
Ability of rock to transport fluid	Very low	Usually, higher
Recovery potential	Relatively low	Relatively moderate
Gas–water contact	Not well defined	Well defined
Occurrences of water	Up dip from hydrocarbon	Downdip from hydrocarbon
Reservoir pressure anomaly	Overpressure reservoirs	Pressure anomalies are relatively few

Source: Zou C. Unconventional Petroleum Geology, Elsevier 2012.

1.6 A LIST OF THE VARIOUS DISCIPLINES THAT MAKE UP PETROLEUM ENGINEERING

There are numerous disciplines involved in petroleum extraction that enable project identification, the generation and evaluation of alternatives, and project success evaluation. A typical process for planning, carrying out, and finishing a project to manufacture hydrocarbons must serve a number of purposes. The workflow involves professionals with expertise from many different fields. For instance, geophysicists and petroleum geologists employ technology to describe the rock of hydrocarbon reservoirs.

1.6.1 Exploration Stage

The exploration stage is the main stage of finding a commercially viable field. An oil or gas field's life cycle begins with the exploration stage, during which undrilled land is analyzed to identify its potential for later commercial development. Exploratory geologists and geophysicists, among others, develop possibilities that are promising for further evaluation during the exploration stage of field development. After these prospects are created, exploration wells are dug to determine whether hydrocarbons are indeed present where the geologists have identified them. A wildcat well is the first exploration well that is drilled in a geologic basin. Predicting viable production potential by subsurface sequestration and formation evaluation is an important function of the professional disciplines. The preparation of environmentally friendly

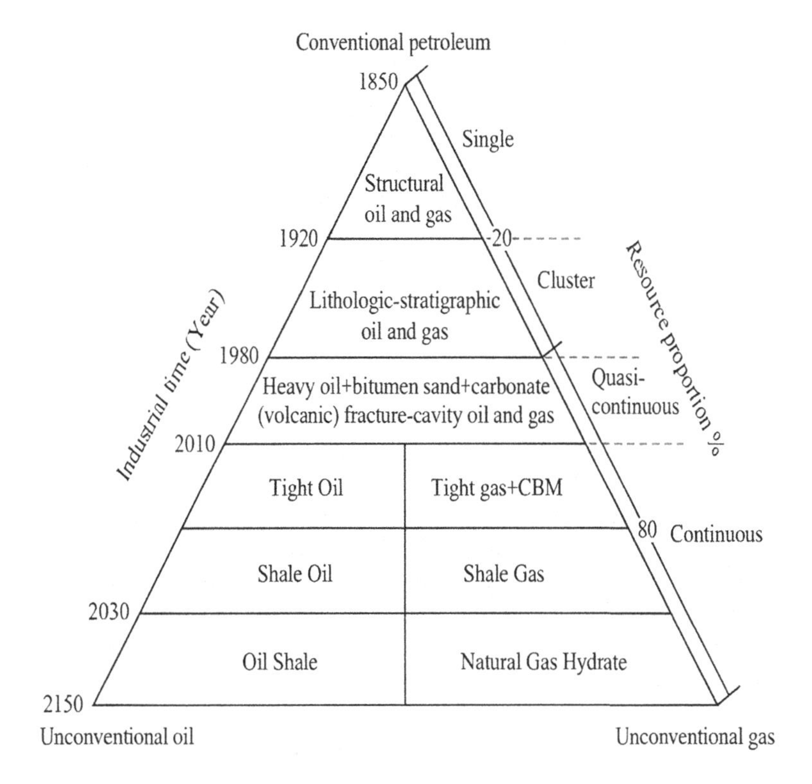

FIGURE 1.3 Petroleum Resource Triangle which is showing different petroleum resources (Conventional and Unconventional) and the beginning year of their exploration/ extraction (adapted from Zou C. Unconventional Petroleum Geology, p. 38. Elsevier; 2017).

Drilling-Mud has advanced primarily to safeguard marine life and maintain biodiversity (Alcheikh and Ghosh, 2017; Agwu and Akpabio, 2018; Patidar et al., 2020). The key disciplines during the exploration stage are:

***Geologists*:** They identify the location of the reservoir deposits by analyzing rock formations. In order to analyze samples for the existence of oil and gas, they gather rock and sediment samples from sites using drilling and other techniques. They work with drilling engineers to develop drilling prospects for exploration wells. Exploration wells are wells which are drilled for the purpose of confirming or disproving the occurrence of a hydrocarbon accumulation. A wildcat well is the first exploration well that is drilled in a geologic basin.

***Geophysicists*:** They employ sophisticated technology to gather information on seismic waves and earthquakes as the waves travel through and around the planet. Geophysicists often concentrate on acquisition, processing, or interpretation within petroleum exploration and production. They give input regarding the surface location of the rig site and also anticipate the geologic cross-section and stratigraphic column for the drilling operation. They anticipate the time required to drill the well.

Reservoir Engineers: Through correct well placement, production output levels, and better oil recovery processes, reservoir engineers aim to maximize oil and gas production. They aid in the establishment of all developed and undeveloped reserves, manage effective reserve systems, assess all drilling and completion efforts for various projects, and work in tandem with various teams to create projects in accordance with production standards. Development of optimal production techniques through reservoir models to simulate fluid behaviour in a reservoir is the key role of a reservoir engineer. They acquire the required data that is essential for the exploration geologist to appraise the well.

1.6.2 The Appraisal Stage

The appraisal stage is closely related to the exploration phase since it seeks to learn more about the reservoir so that decision-makers can decide whether or not to move forward with field development. Appraisal is undertaken to determine the size and characteristics of a petroleum discovery. Prior to the start of commercial hydrocarbon production from a well, appraisal wells are drilled into a discovered formation with good hydrocarbon presence. The cost of this information, which typically includes the purchase of extra seismic data as well as the drilling of appraisal wells, must always be greater than its value. Utilizing computer simulation models of the reservoir, the important reservoir characteristics are identified and ranked according to how they will affect the type and scope of development being considered. The disciplines that are involved in the appraisal phase are:

Reservoir Simulation Engineers: They create simulation models for new fields, determine the quantity of wells needed for optimum well completion, and forecast the hydrocarbon production. Simulation engineers monitor and analyze all surveillance programs for reservoir performance. They create field plans for water injection, provide suggestions for improvements, analyze all test results using industry-standard software, and keep a close watch on field and well performance. They create enhanced oil recovery (EOR) process models for wells that have been producing oil for a while in order to find new ways to increase oil production.

Seismic Interpreters: They use seismic exploration to look into seismological structures by sending equipment and seismometers to various areas across the globe. Seismic interpreters choose appropriate seismic measurement and data processing methods to interpret the data, and write reports. They participate in a variety of activities to promote growth, and work with geoscientists to create development initiatives. They also gather data on rock volume and quality that can be used to estimate the oil or gas yield that is most likely to occur. They also create ground surface maps and cross-sections of the earth's surface.

1.6.3 The Development Stage

Multiple development options are evaluated during the first stage of field development planning. Wells are drilled with the primary goal of producing hydrocarbons throughout the development phase. There is a strong economic incentive (time value of money) for developing the field or reservoir. The field or reservoir life-cycle phase in which the development plan is put into action. The final stages of the development stage focus on creating infrastructure and amenities as well as on operation and maintenance. This is the phase where there is an involvement of multi-discipline technical professionals, namely:

***Maintenance Engineers*:** They carry out maintenance analyses of platform machinery and systems, analyze the nature of the repair job, and provide work order paperwork. They undertake planning and schedule maintenance checks. They are also in charge of determining the source of problems, fixing machinery, and locating specialized parts, fixtures, or fittings.

***Operation Managers*:** With a variety of product lines, including wireline, artificial lift completions, fishing, tubular, liner hangers, and drilling tools, they offer direct monitoring of operation, repair, and maintenance. They oversee the planning and execution of drilling projects for the development of oil and gas fields as well as production, engineering, and drilling activities. To address and fix drilling issues, the operation manager plans and makes modifications or improvements.

***Subsea Engineers*:** Any structural engineering work done below the ocean's surface involves them. They create and install mechanical equipment that is used in the ocean, such as offshore drilling rigs, underwater pipes and pumps, and wellheads. They are also involved in underwater data collection. For subsea development studies, they create field architecture, flow line combinations, and designs. Within the operational subsea facilities, they also provide readiness response tactics as well as pipeline and jumper measures.

1.6.4 Production Stage

The goal of the production stage is to bring the oil and gas to the surface and get them ready for shipping to a refinery or processing facility for further processes. Production stages are identified by the chronological order as primary, secondary, and tertiary production. In the first stage of production, all reservoir fluids are transported to the production wells using natural energy sources. Traditionally, secondary production occurs after primary production and includes the injection of a fluid such as water, or the injection of gas. The injection of water in the reservoir is referred to as water flooding while the injection of a gas is called gas flooding. The disciplines that encompass this stage are:

***Process Engineers*:** Their work entails transferring procedures from the laboratory to the processing plant and focuses on chemical and biological processes in which raw materials undergo change. They design and optimize the operation of equipment and facilities and also advise on the process of safety issues.

***Mechanical Engineers*:** They employ engineering concepts to build effective processes and products, from tiny component designs to enormous plants. They create and oversee project management programs, assess equipment performance, suggest upgrades and new technologies, select materials, aid in equipment operation, and oversee minor renovations and building projects. To ensure smooth start-up and operation of new equipment, mechanical engineers monitor and check the construction progress of assigned projects. They are capable of working on all phases, from design and manufacturing to research and development.

***Rotating Equipment Specialists*:** They ensure that rotating equipment is run and maintained in a way that maximizes its longevity, effectiveness, and operating expenses. They are responsible for solving technical issues and safeguarding the quality of technical maintenance, as well as overhauls of any rotating equipment in the assigned maintenance department. They also participate in modification and improvement projects. Additionally, they assist the manufacturing facilities with the necessary discipline during regular operations as well as interventions such as turnarounds and repairs.

***Production Technicians*:** The flow up of the production tubing, the interface with the surface processing facility, and any potential production constraints are their main concerns. They provide guidance to the field staff about the management of production data and take part in incidence reviews and job safety evaluations. Some of the crucial tasks performed by a production technician include gathering and predicting data on production expense, creating production graphs, and assessing production patterns using computer software. Additionally, they facilitate spill reporting and offer solutions for it.

1.7 ANALYZING RUDIMENTARY ENGINEERING METHODS IN EXPLORATION AND PRODUCTION

Exploration and production procedures use expensive, technologically advanced techniques. Petroleum exploration and production during the prospecting phase involve the use of seismic data interpretation and huge computational data to process the geophysical data as well as subsurface geology evaluation with the aid of gravity and magnetic surveys. Exploratory well drilling is expensive and dangerous. Every choice made during the exploration, drilling, and production phases is weighed against the expected potential economic return on investment. A list of key geophysical techniques is shown in **Table 1.3**.

1.7.1 Role of Geoscientists

Most geoscientists work for companies that utilize various surveys to look for oil and gas. Geologists, geophysicists, and geochemists make up the three primary categories

TABLE 1.3
Brief Details of the Geophysical Methods Used In Oil And Gas Industries. The Measured Physical Properties and Their Uses to Interpret the Subsurface Geological Structures are Also Shown

GEOPHYSICAL TECHNIQUE	PHYSICAL PROPERTY MEASURED	EARTH PROPERTIES INFERRED FROM MEASUREMENTS
Seismic reflection	Travel time to acoustic boundaries, amplitude and velocity	Geological structure, depositional history, faults, rock layers and reservoir size
Gravity	Gravitational attraction, density contrast	Geological structure, spatial variation in rock types such as shale diapers, salt domes
Magnetic	Magnetic field variation, magnetic susceptibility contrast	Geometry of the basement below the sediments, sedimentary cover thickness
Electromagnetic	Changes in electrical conductivity	Shallow near surface lithology changes

Source: Shivaji et al., 2013.

that make up the geoscientists. In order to seek the geological sites where hydrocarbons may be located, each of them focuses on a different component of the task. Geologists are responsible for understanding lithology, geophysicists analyze subsurface structure using gravity, seismic, and magnetic surveys, and geochemists are responsible for understanding subsurface fluids Because they can use their knowledge of earth sciences to understand the underlying conditions and processes operating in sedimentary basins that serve as the "hydrocarbon habitat" for oil and gas deposits, geoscientists are employed by oil exploration and production companies. They work in multidisciplinary teams that include experts from every aspect of oil and gas, such as reservoir engineers, production engineers, environmental analysts, facilities operations, accounting, legal, commercial, and negotiation experts. The challenge that geoscientists face is to interpret and understand the subsurface where hydrocarbons can be present. They do so by studying the regional geology which provides the understanding of the areas which are productive in terms of oil and gas exploration. Geologists and geophysicists predict the presence of oil and gas by the use of a "remote sensing" process which includes gravity and magnetic surveys, rock distribution properties, and geophysical imaging tools to gather data on subsurface characteristics. The data is then subjected to a computing process and advanced modelling approaches with 3-D visualization of the data for a better understanding of the subsurface. In addition to surveys, remote sensing, and modelling techniques, structural geology and stratigraphy are two more factors that are heavily utilized in the oil and gas sector. Understanding the underlying deformation mechanisms that led to the creation of hydrocarbon traps is provided by structural geology. Stratigraphy is the study of how the different rock layers originated and deposited in succession. More recently, Artificial intelligence (AI) and machine learning (ML) techniques are being used in data management at various stages of hydrocarbon exploration to development

1.7.2 Methods of Exploration

An exhaustive investigation of the petroleum system is necessary during the exploration phase. The term "petroleum system" refers to the entire set of components and operations that make up petroleum geology, including source rock, reservoir rock, migratory channel, trap, and seal. Strong circumstances for substantial oil and gas discoveries are created by the source rock's maturation period and the hydrocarbons' efficient migration into the sealed reservoir (Reis and Pimentel, 2021). Following the completion of the exploration project, where geologists and geophysicists conduct gravity, magnetic, and seismic surveys to understand the subsurface as well as the environmental risks associated with the procedure, the decision to drill on the appropriate geological site is made. A wildcat well or exploratory well is one that is drilled with the intention of finding a new oil or gas reservoir. The well is known as the discovery well in the event of discovery. Evaluation of volumes and well productivity becomes a priority following the discovery. The quantity of wells that must be drilled, the processes necessary to recover the hydrocarbons already present, the techniques used to collect the fluids, and the infrastructure and installations required for the production processes are all included in a plan for the field development.

1.7.3 Methods of Production

The production stage is reliant on numerous contracts and agreements with landowners or the concession's authority. An area of land or the ocean floor may be specified in a contract to be explored during a specific period of time known as contract time. An exploratory concession is granted to the multinational firm, and during the period of the contract, the corporation is in charge of paying all drilling and exploration costs (Reis and Pimentel, 2021). A brief life cycle of an oil gas field is illustrated in Figure 1.4.

If commercially significant quantities of oil and gas are discovered, the company receives a predetermined portion of the gross oil and production—known as "cost oil"—to sell and recover the costs of the exploration, drilling, and production. There are four stages in the production of hydrocarbons. The first one involves primary recovery, in which the production is driven by the natural energy stored in the reservoir; 5–15% of the reservoir's oil can be extracted. The secondary recovery mechanism, sometimes referred to as "pressure maintenance", is a part of the second phase. This forces the crude oil to the base, where it may be pumped, while the pressure of the reservoir is maintained by pumping water or natural gas into the well; 35–45% of the oil can now be extracted from the reservoir at this point. The third phase includes the EOR technique, which extracts 5–10% of the oil. The final phase is the decline phase, during which production gradually decreases until the reservoir is depleted.

1.8 INTERPRETATION OF CROSS-PLOTS

The finding of new hydrocarbon reserves and the recovery from current deposits are becoming more difficult due to a lack of knowledge about the reservoir and the

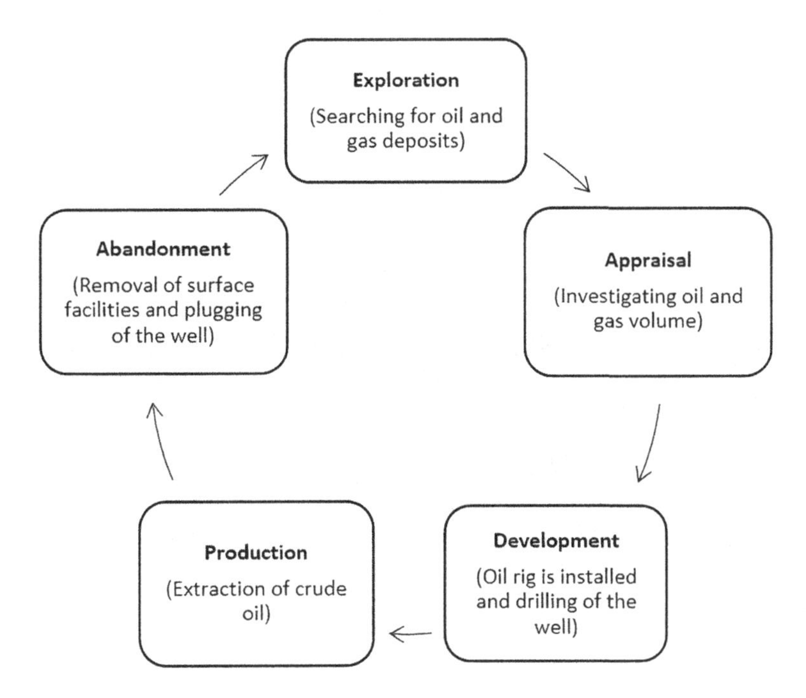

FIGURE 1.4 Life cycle of an oil gas field.

subsoil. It is possible to tell whether or not a hydrocarbon reservoir can be used for production by understanding its many properties, such as thickness, porosity, permeability, etc. Due to their assistance in the volumetric analysis of the reservoir, these properties are crucial. In-depth petrophysical study is required to maximize reservoir development and production because the reservoir features and lithofacies may occasionally be difficult to anticipate due to poor core sample quality. By employing well log cross-plots, this issue can be resolved by determining the reservoir parameters (and various petrophysical research). Cross-plots help to understand lithology by providing comparison plots between two or more reservoir metrics. We may identify the different lithologies (Figure 1.5), such as sandstone, shale, saline aquifers, etc., by plotting cross-plots of well logs (Figure 1.5) such as resistivity vs. gamma, density vs gamma, etc. For example, cross-plots of sonic versus density logs are frequently used in the analysis of shaly sands. For carbonates, the density vs. neutron cross-plot is frequently utilized.

The cross-plot analysis can also help in identifying geological sites for the long-term storage of enormous amounts of energy, which is used seasonally in the form of CO_2 sequestration. CO_2, when stored as an energy storage medium in its supercritical condition in reservoirs under the caprocks (at a particular depth), generates more energy than the energy that is stored. Large volumes of CO_2 can be held in deep saline aquifers, or depleted oil/gas fields (essentially permeable sedimentary reservoirs), and its vertical leaking can be controlled by overlying low permeability cap rock. The implications of the cross-plotting technique can also help in the

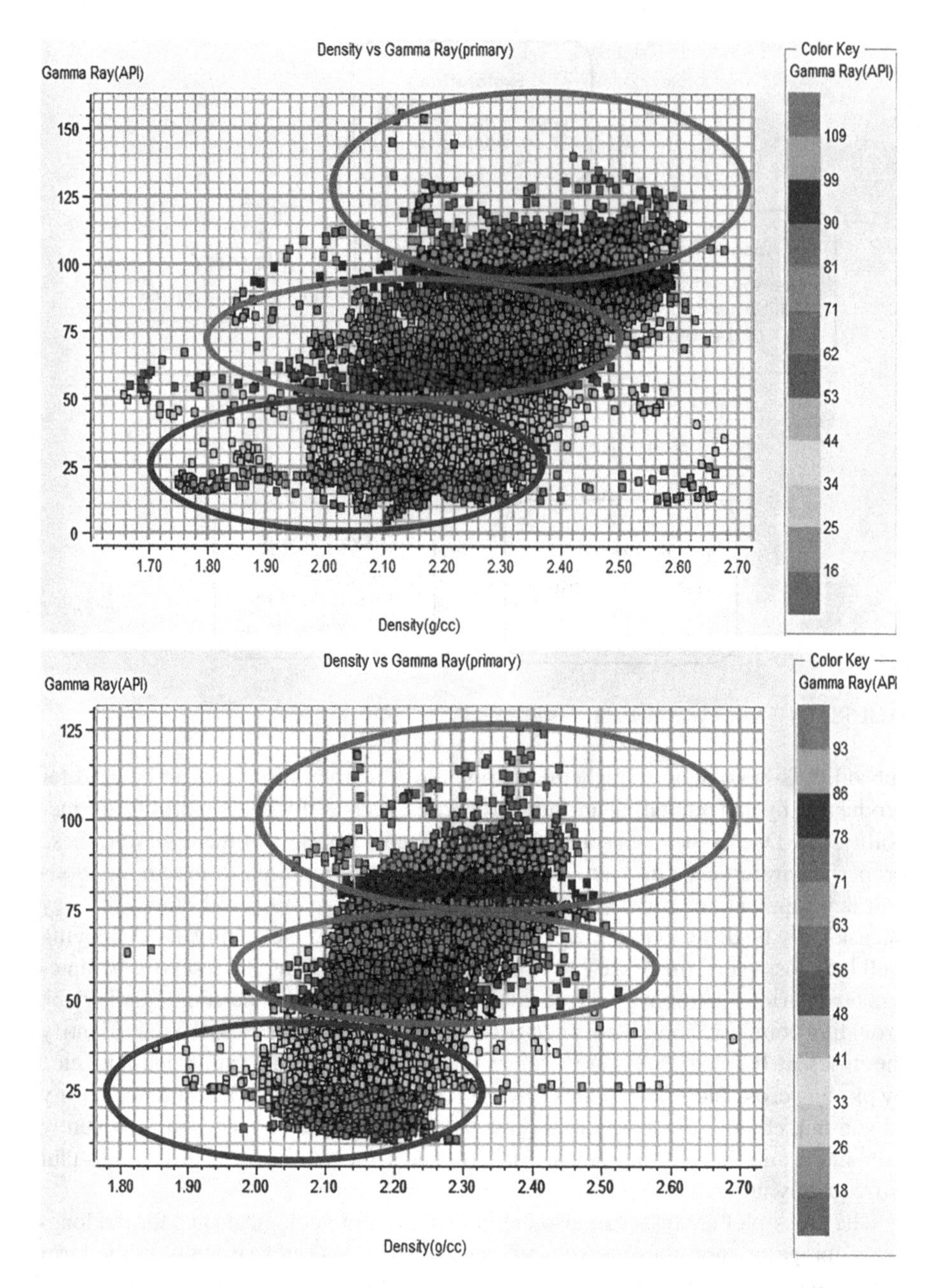

FIGURE 1.5 Lithological identification (petrophysical analysis) using cross-plot between gamma ray and density (adapted from Michael et al., 2018).

identification of different geological lithofacies for CO_2 sequestration. With reference to Figure 1.5, the green fill colour aids in the identification of low gamma ray values. Sands can be detected by low gamma ray levels. The lowest cluster (blue ellipse) correlates to the sand zone, while the middle cluster (green ellipse) refers to a unit made up of an intercalation of sand and shale. The greater cluster of high gamma ray and high-density values, represented by the red ellipse, shows the well's more shaly interval (Figure 1.5).

ACKNOWLEDGEMENT

The support received from UPES, Dehradun, is acknowledged by the authors.

REFERENCES

Agwu, O. E., and Akpabio, J. U. (2018) Using agro-waste materials as possible filter loss control agents in drilling muds: A review. *Journal of Petroleum Science and Engineering*, 163, 185–198. https://doi.org/10.1016/j.petrol.2018.01.009

Alcheikh, I. M., and Ghosh, B. (2017) A comprehensive review on the advancement of non-damaging drilling fluids. *International Journal of Petrochemistry and Research*, 1, 61–72. https://doi.org/10.18689/ijpr-1000111

Dalvi, S. (2015) *Fundamentals of Oil & Gas Industry for Beginners*. ISBN 978-9352064199

Dasgupta, S. N., and Aminzadeh, F. (2013) *Geophysics for Petroleum Engineers*. Elsevier. ISBN: 978-0-444-50662-7

Fanchi, J. R., and Christiansen, R. L. (2017) *Introduction to Petroleum Engineering*. Wiley. ISBN 9781119193449

Gao, Z. (1998) *Environmental Regulation of Oil and Gas*. Kluwer Law International, p. 8. ISBN 9789041107268

Hassan, A. (2013) Review of the global oil and gas industry: A concise journey from ancient time to modern world. *Petroleum Technology Development Journal*, 3(2), 123–141.

Herbert, A. (1942) *The Golden Flood, An Informal History of America's First Oil Field*. ASIN: B0007DOC4U, A.A. Knopf; First Edition (January 1, 1942).

Joshi, D., Patidar, A. K., Mishra, A., Mishra, A., et al. (2021) Prediction of sonic log and correlation of lithology by comparing geophysical well log data using machine learning principles. *GeoJournal* (Springer). https://doi.org/10.1007/s10708-021-10502-6

Kilrani, N., Prajapati, P., and Patidar, A. K. (2021) Contrasting machine learning regression algorithms used for the estimation of permeability from well log data. *Arabian Journal of Geoscience* (Springer), 14, 2070. https://doi.org/10.1007/s12517-021-08390-8

Mishra, A., Sharma, A., and Patidar, A. K. (2022) Evaluation and development of a predictive model for geophysical well log data analysis and reservoir characterization: Machine learning applications to lithology prediction. *Natural Resources Research*, 31, 3195–3222. https://doi.org/10.1007/s11053-022-10121-z

Mishra, M., and Patidar, A. K. (2022) Post-drill geophysical characterization of two deep-water wells of Cauvery Basin, East Coast of India. *Journal of Petroleum Exploration and Production Technology*. https://doi.org/10.1007/s13202-022-01550-w

Ohakwere-Eze, M., Igboekwe, M., and Chukwu, G. (2018) Petrophysical evaluation and lithology delineation using cross-plot analysis from some onshore wells in the Nigerian-delta, West Africa. *International Journal of Advanced Geosciences*, 6(1), 99–107. https://doi.org/10.14419/ijag.v6i1.9601

Patidar, A. K., Joshi, D., Dristant, U., and Choudhury, T. (2022) A review of tracer testing techniques in porous media specially attributed to the oil and gas industry. *Journal of Petroleum Exploration and Production Technology* (Springer). https://doi.org/10.1007/s13202-022-01526-w

Patidar, A. K., Sharma, A., and Joshi, D. (2020) Formulation of cellulose using groundnut husk as an environment-friendly fluid loss retarder additive and rheological modifier comparable to PAC for WBM. *Journal of Petroleum Exploration and Production Technology*, 10, 3449–3466. https://doi.org/10.1007/s13202-020-00984-4

Rakesh, M., Rakesh, P. K., Kumar, B., Chowdhury, S., and Patidar, A. K. (2021) Numerical simulation of gravity driven turbidity currents using computational fluid dynamics. *Materials Today Proceedings*. https://doi.org/10.1016/j.matpr.2021.09.238

Reis, R. P., and Pimentel, N. (2021) *Exploration and Production of Petroleum*. Springer Nature Switzerland AG, pp. 1–9. https://doi.org/10.1007/978-3-319-71064-8_45-1

Sahin, A. (2013) *Unconventional Natural Gas Potential in Saudi Arabia*. https://doi.org/10.2118/164364-MS

Zou, C. (2017) *Unconventional Petroleum Geology*. Elsevier. ISBN-978-0-12-812234-1

2 IT Technologies Impacting the Petroleum Sector

K.T. Igulu, Z.P. Piah, and Thipendra P Singh

2.1 INTRODUCTION

Oil is the primary non-renewable energy source currently "fuelling" the global economy. While the rising cost of international crude oil has certainly bolstered efforts to develop renewable energy sources, their share of global energy consumption remains negligible despite these efforts. The evolution of technology has an impact on all aspects of the energy market and all regions of the globe. The extremely competitive and global petroleum industry encourages the global dissemination of technology. The advancement of technology contributes to the expansion and growth of the economy. Unavoidable, but notoriously difficult to predict, is the development of new technologies (Ike et al., 2013).

A large portion of the economy relies on IT to either boost productivity or decrease expenses With sustained high oil prices expected, it's worth asking if there are ways to improve production and distribution of this precious energy commodity, such as through the application of cutting-edge information and communication technologies. Can the use of information technology be leveraged to improve crude oil and product output?

As information and communication technologies (ICTs) and cutting-edge petroleum technologies converge, they present opportunities for boosting economic performance at every stage of the oil supply chain. As a result of these developments, both the upstream (oil exploration and production) and the downstream (refining and distribution) sectors of the oil industry are impacted (transportation, refining of crude oil, and distribution of oil products). It has been hypothesized that incorporating ICTs and related technologies into upstream operations could increase proven crude oil reserves, speed up the extraction of crude oil from existing wells, and provide additional tools for locating new wells.

Differentiating between basic ICTs such as email and the Internet, and more complex solutions such as integrated data networks or censor devices measuring the drilling and extraction processes is crucial. When it comes to basic ICT solutions, both international oil companies (IOCs) and national oil companies (NOCs) in developing countries rely heavily on the increasingly standardized options available to them.

DOI: 10.1201/9781003357872-2

Although only the largest international oil companies have fully adopted the most cutting-edge information and communications technology solutions, this is quickly changing. It is estimated that the oil industry spends over $10 billion annually on ICTs for their upstream, downstream, and petrochemical operations. This includes companies such as BP, Conoco-Phillips, Chevron, Exxon-Mobil, Shell, ENI, and Total, which are all descendants of five of the infamous seven sisters. Effects of information technology on the petroleum industry are discussed here (ESRI, 2007).

The number of computers and other ICT tools available to each oil industry employee is one indicator of the industry's growing reliance on these tools. Increased performance in measuring and computing capabilities of new oil equipm.ent indicates the increasing information-intensive nature of the equipment and its integration with ICTs within a more holistic business using ICTs, which could contribute to the extension of proven reserves beyond current levels and the acceleration of the extraction of crude oil from existing wells and products from crude oil. An increase in ICT use may also be implied by the competitive pressures exerted by manufacturers of alternative energy sources.

A report by the UN presents IT in the petroleum value chain. Figure 2.1 illustrates this.

2.2 GIS AND REMOTE SENSING

A geographic information system (GIS) allows users to analyze data in a geographic context and make more informed decisions (Smith and Hania, 2000). The goal of a geographic information system is to make it easier to find the information you need when you need it using maps. Layers of information can be added to a real-world basemap via a thematic map's table of contents. Using a basemap of Eugene, Oregon

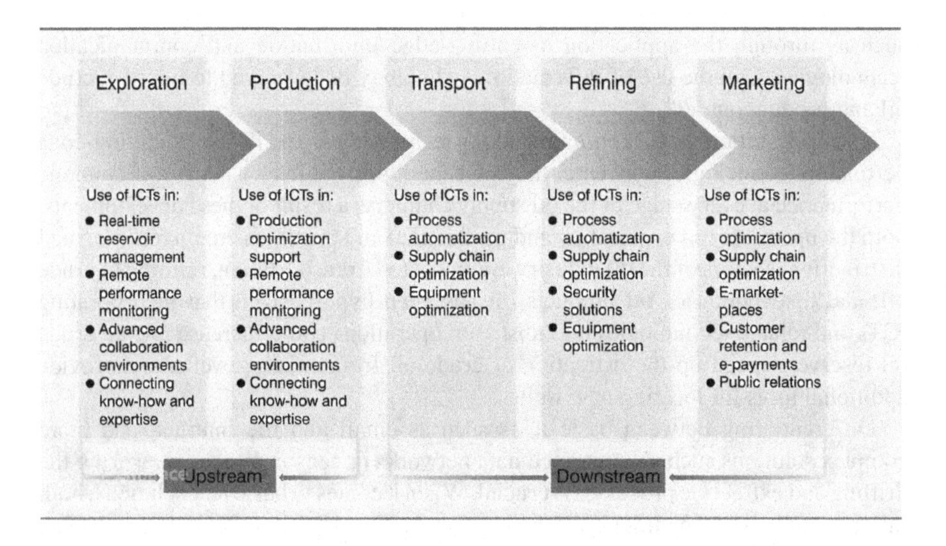

FIGURE 2.1 Uses of IT technologies in the petroleum value chain (UN, 2006).

and selected datasets from the United States Census Bureau, a social scientist could produce a map displaying demographic information about the city's residents, including age, education level, and employment status. Due to GIS's flexibility in accommodating a wide range of dataset combinations, it can be applied to a wide range of disciplines, from archaeology to zoology (Bylov, 2013).

A powerful GIS application can consolidate and analyze geographical information from numerous sources. Governments in many nations also frequently release publicly available GIS datasets, meaning that analysts can get their hands on a wealth of previously undisclosed geographic information. Many GIS software packages include access to pre-built map file databases, while additional databases can be purchased from private companies or obtained from public sources. On-field global positioning systems (GPS) collects data by assigning a feature such as a pump station on a geographic coordinate (latitude and longitude). Geographic information system maps can be manipulated by the user. Viewers of a GIS map on a computer can pan the image in any direction, zoom in or out, and alter the type of data displayed. They have control over whether or not roads are displayed, how many roads they see, and how those roads look. The user is then given the option to customize the information displayed alongside these roads. This includes things such as storm drains, gas lines, rare plants, and hospitals. In order to keep tabs on storms and anticipate erosion patterns, some GIS programs are programmed to carry out complex calculations. Common tasks, such as checking an address, can easily incorporate GIS applications.

Geographic information systems provide users with a geographical advantage in a wide range of activities, from day-to-day work tasks to scientific exploration of the complexities of our world, allowing them to improve their productivity, awareness, and responsiveness as global citizens (Smith and Hania, 2000).

The future of the petroleum industry hinges on finding new sources of petroleum before the competition does. Potential sites for oil extraction can be mapped and analyzed using a GIS. Satellite images, digital aerial photomosaics, seismic surveys, surface geology studies, interpretations, and images of the subsurface and cross sections, well locations, and details about the existing infrastructure are just some of the types of data that need to be analyzed. These data points can be plotted on a map using a GIS, where they can be viewed, overlaid, and manipulated to shed light on the data's significance. In order to apply appropriate geographic analysis across the enterprise, modern GIS technology enables efficient management of the spatial components of typical petroleum "business objects" within the corporate database. Leases, oil fields, pipelines, ecological issues, buildings, and stores all fall under this category (He and Wu, 2003).

2.2.1 Case Study of GIS in the Petroleum Industry 1: OMV Enterprise GIS

OMV is utilizing the ESRI Multinational Enterprise Agreement as it is the largest oil and gas company in central and eastern Europe. The Vienna, Austria-based company OMV is rolling out a global enterprise GIS solution from ESRI. OMV is widely recognized as a pioneer in the oil and gas industry. Many of the GIS capabilities are applicable to the oil and gas sector (E&P). With the help of its ArcGIS enterprise licence, the company is making the most of GIS by creating analysis tools, reducing data redundancy, increasing access to data across applications, and improving the overall E&P workflow (ESRI, 2007).

OMV uses GIS for a variety of purposes across their exploration and production operations in 17 countries. GIS is able to quickly and efficiently store, retrieve, analyze, and visualize spatial data by integrating geographic information within a database management system. The exceptional mapping, querying, and analysis capabilities of ArcGIS have not gone unnoticed by the providers of specialized E&P applications. Applications are increasingly including ArcGIS interfaces as a means of data exchange or for performing specialized tasks.

Making sure the right data is accessible to the right people at the right time and in the right places is a cornerstone of effective data management. For this task, using a GIS is strongly suggested. Accurate environmental knowledge is essential when making decisions about where to drill a well, how to avoid collisions along a winding well path, or where to lay a pipeline. Information management is often a challenging task. ArcGIS is helping the petroleum industry deal with its global scope, the ever-increasing volume and variety of data to process, and the complexity of the procedures involved in finding new petroleum reserves (ESRI, 2007).

GIS facilitates access to well-organized data, which is essential for making sound decisions. E&P managers and engineers at OMV use ArcGIS 9.1 to generate maps for geological field studies, seismic survey planning, management production overviews, and regulatory reports. The central data warehouse of OMV, regional data warehouses, and third-party providers such as IHS Energy are among the locations where data is collected. Contract areas, blocks, wells, and well-related information, such as well formation tops and deviations, are imported into GIS from the corporate databases of OMV. Seismic data navigation maps created with GIS provide access to metadata describing seismic traces and processing parameters. Welcoming additions include satellite data, orthophotos, digital elevation models, and the production database (ESRI, 2007).

Access to the company's multiuser geographic database and distribution of spatial data to ArcGIS Desktop users and Internet clients via ArcIMS are both critical functions of ArcSDE within OMV's enterprise system. Feature classes are used to categorize information and connect it to context. When a business adopts the idea of an enterprise spatial database, it eliminates the need for multiple data repositories and paves the way for the delivery of timely, reliable, and uniform thematic maps that can be shared via a web-based Internet mapping service within the business and with outside stakeholders.

Experts in data management, information systems, and the practical applications of geology and geophysics make up OMV's Knowledge Management Group. The team has launched an ambitious multi-project work plan. Its primary objective is to make high-quality data, information, and knowledge accessible to professionals in a wide range of E&P fields and end users whenever and wherever they may be (ESRI, 2007).

Unstructured data, such as reports in a document management system; structured data, such as that found in databases; and spatial data, such as that found in E&P

applications and the GIS are the three main types of data used in the petroleum industry. Previous attempts at connecting these data types failed. It is the goal of the current projects to integrate all of this information. OMV's growing reliance on ArcGIS Desktop prompted the company to create a networked enterprise GIS to further enhance the software and better tap into OMV's internal data. A business GIS integrates data from multiple departments into a single, shared database. As a result, GIS is now connected to the rest of the company's IT infrastructure.

As part of its preparations to migrate its corporate database into the Public Petroleum Data Model, OMV E&P intends to implement scripts to automatically extract concession maps, seismic coordinates, well data, and production charts from the database to spatially enable these data.

OMV has access to two- and three-dimensional views of seismic data through a GIS-enabled website, as shown in the two images below. The ArcIMS viewer, depicted in the image below, enables users to quickly access E&P data online and generate planning-friendly maps.

2.2.2 GIS Case Study 2: Assessment of Hurricane Effects on Gulf of Mexico Oil and Gas Production Predictions

2004 and 2005 hurricanes in the Gulf of Mexico had a significant impact on the oil and gas industry. The Minerals Management Service estimates that 105,889,263 barrels of oil production in the Gulf of Mexico were halted between August 26 and December 22, 2005 due to damage from Hurricanes Katrina, Rita, and Wilma (MMS). This amounts to about 19% of the region's total annual oil output. Katrina and Rita destroyed 111 buildings and forced the relocation of 19 oil rigs, resulting in $11.7 billion in damage. A new GIS analysis program was tested during the crisis (ESRI, 2007).

Earth Science Associates (ESA) analyzed the risks posed by Hurricanes Katrina and Rita to the oil and gas industry and shared its findings with its clients and MMS. GIS software was used to compile both the hurricane path and intensity projections from the NWS and the oil and gas facility data from the MMS. Access to these data is crucial for conducting accurate risk assessments and developing effective recovery strategies.

By combining data from the MMS and ESA, we have a rich resource for generating the correlations and overlays necessary for hurricane risk assessment. The ESA database contains details on pipeline networks and 50,000 wells in the Gulf Coast. Very few of ESA's current clients produce the vast majority of the oil and gas extracted from the Gulf of Mexico. GIS spatially represented predicted hurricane corridors and vulnerable assets to better prepare their clients for impending hurricane disasters.

2.3 IMAGE PROCESSING

The term "image processing" is used to describe the various methods that can be used on an image to either enhance it or extract useful information from it. "Image processing" is a subset of signal processing that takes an image as input and returns

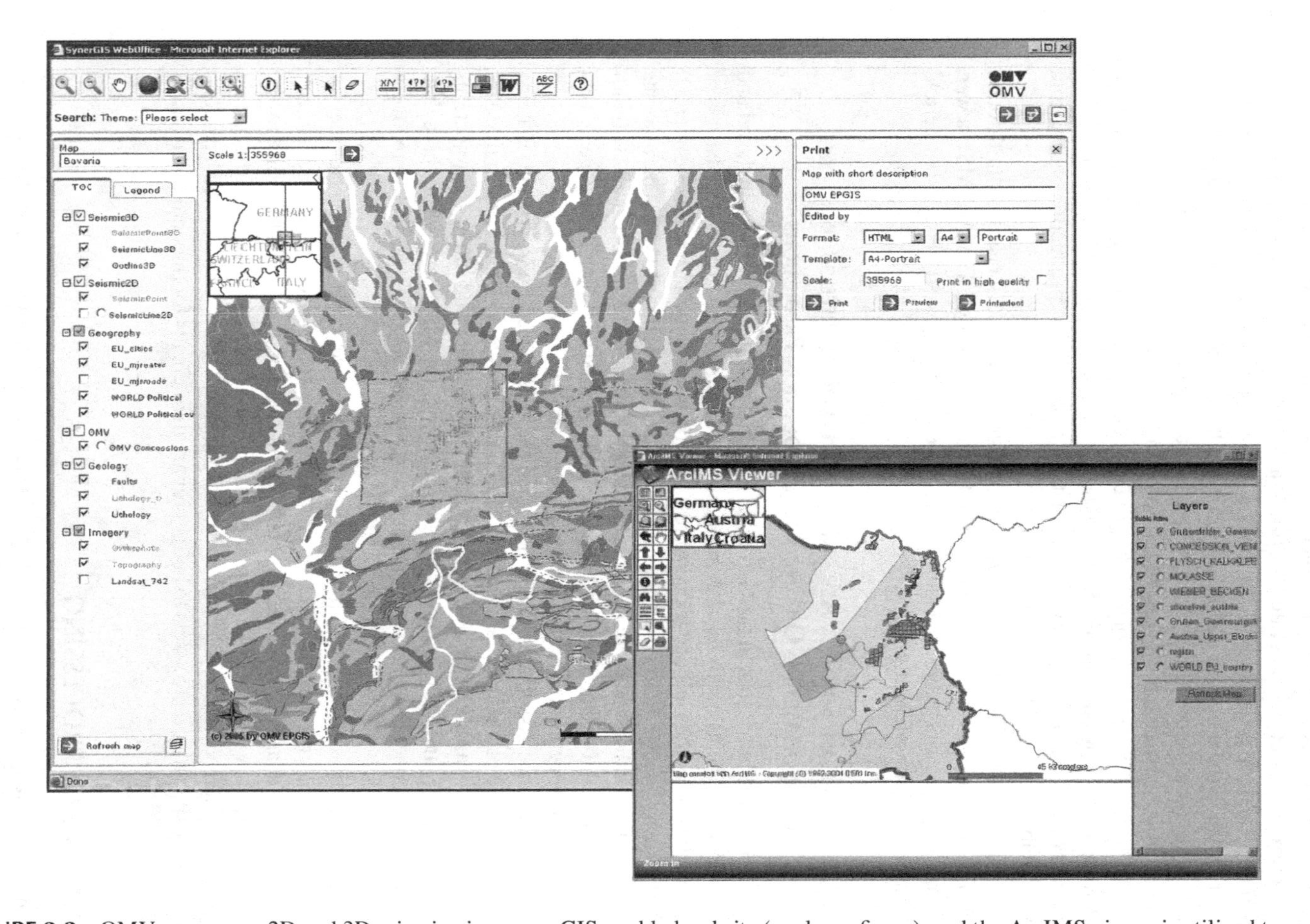

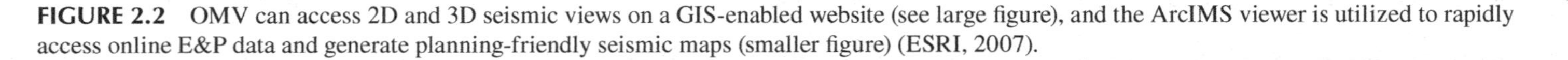
FIGURE 2.2 OMV can access 2D and 3D seismic views on a GIS-enabled website (see large figure), and the ArcIMS viewer is utilized to rapidly access online E&P data and generate planning-friendly seismic maps (smaller figure) (ESRI, 2007).

the same image or a set of features or properties similar to the input. One of the many rapidly developing technologies currently is image processing. A lot of work is being done in this area in engineering and computer science as well.

There are three main phases in image processing:

i. The image acquisition
ii. The image analysis
iii. Output phase/ report generation phase.

In the field of image processing, both analog and digital techniques are applicable. Examples of tangible media that can benefit from analog image processing are prints and photographs. Image analysts must apply a unique set of interpretation principles for each visual analysis technique. When it comes to manipulating digital images, computer-aided digital image processing techniques are invaluable. When dealing with data in digital form, there are three main steps: pre-processing, enhancement, and display, information extraction.

2.4 SCADA AND TELEMETRY

Unlike in other settings, pipelines require a special kind of regulation. While distributed control systems (DCS) are typically built for fast, unrestricted communications within a plant, supervisory control, and data acquisition (SCADA) systems are designed for low-speed, limited-bandwidth communications over long distances. SCADA systems typically work with small data packets. Instead of keeping data such as controller IDs in the field controller itself, they are kept in a centralized database. Controllers save operational data for infrequent polling, and controllers are designed to report by exception for changes in status, reducing the need for constant communication. In order to reestablish history in the event of a communication link failure, controllers typically record the sequence of events and require confirmation of each link with each report. The data acquisition module of a SCADA system is in charge of gathering field data and saving it to a database from which it can be accessed by other modules. SCADA systems can be extremely comprehensive, collecting data from tens of thousands of points multiple times per day. Pipeline control systems require the collection of massive amounts of data at each instant as a standard requirement. Humans interact with the SCADA system, which is responsible for supervisory control, primarily through the HMI. Here, a user can access graphical representations of the current values of the point database and issue commands to the appropriate field devices.

When it comes to controlling and monitoring massive industrial processes that span multiple locations and long distances, a system known as supervisory control and data acquisition (Scada) is indispensable. The typical components of a SCADA system are as follows:

i. An HMI is a device that relays data about a process to a human operator, letting that person keep tabs on and adjust the operation of the process.

ii. A computerized monitoring and control system that takes in process information and issues directives.
iii. Connecting sensors to remote terminal units (RTUs), which digitize sensor signals and transmit them to a central control system.
iv. Field devices typically consist of PLCs rather than RTUs due to the former's greater adaptability.
v. The network of transmitters and receivers that transmit data between the central monitoring system and the outlying terminals.

More optimization, safety, and efficiency can be achieved for the pipeline owner via the SCADA system's. Execution of a wide variety of applications are made possible through automation and some human control. Batch scheduling, batch tracking, composition tracking, drag reducing agent (DRA) tracking, product distribution, inventory analysis, allocation, nomination, pig scheduling and tracking, training simulator, look-ahead, survival time, predictive modelling, pump optimization (between stations), compressor optimization (between stations), scenario study, pipeline efficiency, are only few of the operations which are achieved with the help of SCADA and GIS (Geographic Information Systems) technologies. and more

2.4.1 Telemetry

The term "telemetry" refers to the practice of sending data collected by a measuring device over a long distance using a variety of communication channels (such as radio, telephone, etc.). Telemetry systems in the oil and gas industry boost productivity in dispersed field operations and mission-critical infrastructures. To meet the challenges of maximizing field production and relaying secure, reliable data from remote assets to the corporate enterprise, these systems aim to optimize data transfer across remote communication networks. Because of its dependability, flexibility, and low cost, telemetry is widely used in the oil and gas industry for remote asset monitoring. Field technicians and engineers can take advantage of the system's simplicity, scalability, and improved safety.

Oil and gas production often takes place in inaccessible locations where it would be too expensive to run wired or ground-based wireless networks. Production wells, along with the pumps and recovery facilities they employ, are crucial to the smooth running of oil and gas production sites, as they are responsible for extracting the raw materials before sending them on to transmission pipelines. If these parts break down, it could cost a lot of money. Telemetry systems, however, can provide a precise diagnosis of the problem and significantly improve response time.

When it comes to remote system monitoring, measurement, and control, Otis Link telemetry offers dependable, hassle-free, web-based solutions. Our products are built to last in even the most demanding conditions, making them ideal for use in the oil and gas industry. The Otis Link telemetry system features field units, communication, and web-based software. Not only are our products wireless and intrinsically safe, but they also use very little power. Otis Link telemetry systems can be easily integrated with existing detection products in the field, optimizing the life cycle of remote assets while lowering costs thanks to the availability of multiple communication options.

2.5 GEOLOGICAL AND GEOPHYSICAL PARAMETERS

Geophysical parameters, viewed from the sidelines, are techniques that make use of the Earth's inherent fields, such as gravity and magnetism. For example, seismic reflection is an example of a technique that needs a source of artificially generated energy in order to function. It is the job of geophysicists to detect the presence of underground bodies or structures and then measure their dimensions, velocity, porosity, and other physical properties in order to learn more about them.

2.6 INTRODUCTION TO ANN AND AUTOMATION

The term "artificial neural network" (ANN) refers to "information-processing systems that simulate biological neural networks" (Mohaghegh, 2000). Each living thing is made up of cells. Neurons are a type of nerve cell. The cell body, axon, and dendrites of a typical neuron are depicted in Figure 2.3. Data from the cell body is transported by dendrites. What was once an output from the cell body is now an input to another neuron thanks to the axon's journey (Mohaghegh, 2000).

It's estimated that the brain contains between 10 and 500 billion neurons (Rumelhart and McClelland, 1986). There are 500 neural networks in each subsection

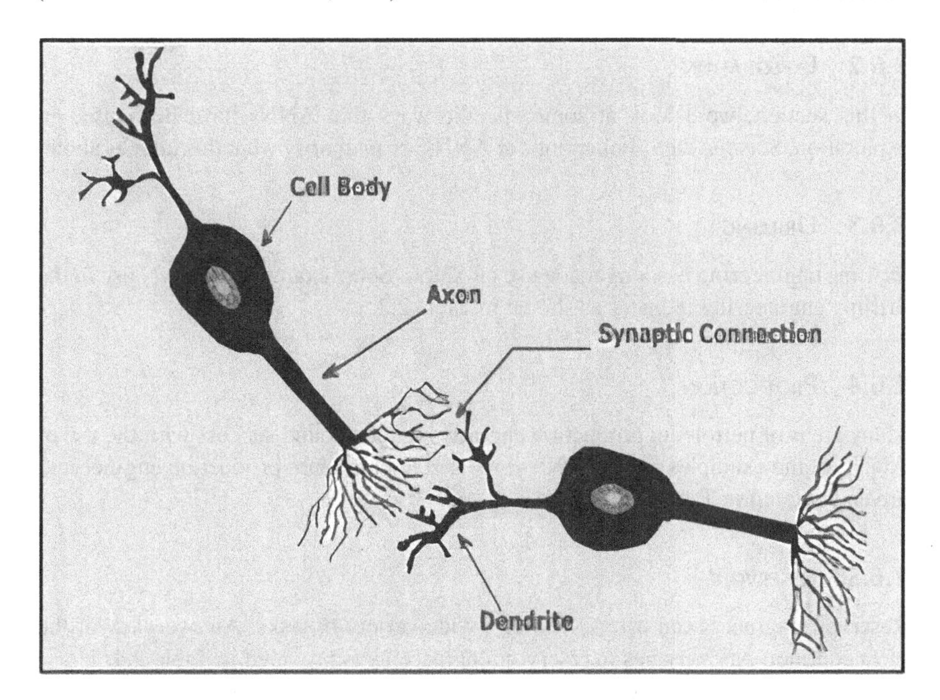

FIGURE 2.3 Two Bipolar Neurons (after Mohaghegh, 2000).

(Stubbs, 1988). One hundred thousand neurons in each neural network are connected to thousands more (Mohaghegh, 2000). In a nutshell, this framework is responsible for explaining why people act the way they do. Even the most fundamental human actions, such as walking, moving one's hands, and grabbing a cup of coffee, require extremely complex calculations that a computer simply cannot handle. Despite computers being faster than human brains (the cycles of computer chip are measured in nanoseconds), the human brain is capable of performing more complex tasks due to the intricate neuronal structures it contains.

ANNs simulate the above biological process. ANNs are based on the following assumptions (Mohaghegh, 2000):

i. Neurons process information.
ii. Neuronal connections let information pass.
iii. Link weights vary.
iv. Neurons apply an action function to inputs to determine outputs.

2.6.1 ANN in the Petroleum Industry

The petroleum industry has made use of ANNs to find answers to difficult problems. ANN applications were divided into four groups after reviewing papers and case studies found in the petroleum literature: exploration, drilling, production, and reservoir (Alkinani et al., 2019).

2.6.2 Exploration

In this section, we'll look at some specific ways that ANNs have been used in exploration. Seismic data applications of ANNs are primarily what this table is about.

2.6.3 Drilling

Drilling engineering has long made use of ANNs. Some examples of ANN use in the drilling engineering industry as shown in Table 2.2.

2.6.4 Production

Many areas of petroleum production engineering have found success with the use of ANNs. Some examples of how ANNs are used in petroleum production engineering are enumerated in Table 2.3.

2.6.5 Reservoir

Reservoir engineers can use ANNs for a wide variety of tasks. An overview of the most common ANN usages in reservoir engineering is provided in Table 2.4.

Finally, ANNs are a powerful tool that can be used to address problems that are challenging to model analytically. There have been many successful applications of

TABLE 2.1
Application of ANNs in Exploration

Author(s)	Application	Notes
Guo et al. (1992)	Feature recognition	Used ANNs to extract structural lineaments and lithologic information from seismic data
Hansen (1993)	Primary reflection identifications	Used ANNs to successfully identify the primary reflection from seismic data
Karrenbach et al. (2000)	Seismic data processing	Used ANNs for seismic data processing.
Fogg (2000)	Petro-seismic characterization	Used ANNs for petro-seismic characterization of a 96,000 trace 3D seismic migrated volume
Xiangjun et al. (2000)	Hydrocarbon prediction	Used ANNs to predict hydrocarbon and presented a case history from DaQing field.
Sun et al. (2000); Russell et al. (2002)	Amplitude variation with offset (AVO)	Used ANNs to solve the interpretation problem associated with AVO since it is hard to distinguish between wet sand and gas sand
Aminzadeh & deGroot (2005)	Object detection	Used ANNs to detect several seismic objects using seismic data
Huang et al. (2006)	Parameters determinations and seismic pattern detection	Used ANNs to detect line pattern of the direct wave and hyperbola pattern of reflection wave in a seismogram
Kononov et al. (2007)	Travel time computation	Used ANNs to compute travel times for a complete 3D volume model.
Huang & Yang (2015)	Seismic velocity picking	Used ANNs to for velocity picking in the time-velocity semblance image of seismic data
Clifford & Aminzadeh (2011)	Gas detection	Used ANNs to detect gas from absorption attributes and amplitude in Grand Bay field
Aminzadeh et al. (2011)	Micro-seismic	Used ANNs to auto pick micro-seismic earthquake data
Verma et al. (2012)	Mapping	Used ANNs to map high frackability and high total organic content zones in the Barnett Shale
Hami-Eddine et al. (2015)	Amplitude variation with angle of incidence (AVA) prediction	Used ANNs to predict AVA to help to evaluate the comparative risk between prospects for ranking purposes.
Refunjol et al. (2016)	Identifying unconventional potential	Used ANNs to identify the unconventional potential using seismic inversion in the Eagle Ford
Ross (2017)	Improve resolution and clarity of seismic data	Used ANNs to improve resolution and clarity of seismic data in the tight sand that has lower porosity, higher bulk density, and velocity. Also, used a practical example form the Permian Basin

Source: Alkinani et al., 2019

TABLE 2.2
Use of Neural Networks in Oil Exploration

Author(s)	Application	Notes
Arehart (1990)	Drill bit diagnosis	While drilling, artificial neural networks were used to assess the drill bit's grade (wear).
Bilgesu et al. (2001)	Drill bit selection	Non-linear, multi-input, multi-output drilling systems were modelled using ANNs.
Ozbaoglu et al. (2002)	Bed height for horizontal wells	Bed heights in wellbores with varying degrees of inclination were predicted using ANNs.
Fruhwirth et al. (2006); Wang and Salehi (2015)	Drilling hydraulics optimization and prediction	Drilling hydraulics optimization and prediction using artificial neural networks.
Dashevskiy et al. (1999)	Real-time drilling dynamic	Non-linear, multi-input, multi-output drilling systems were modelled using ANNs.
Moran et al. (2010); Al-AbdulJabbar et al. (2018a)	Rate of penetration (ROP) prediction	To improve drill time estimations, ANNs were used to predict ROP.
Gidh et al. (2012)	Bit wear prediction	Improved ROP through the use of artificial neural networks for predicting and managing bit wear.
Lind and Kabirova (2014)	Drilling troubles prediction	Utilized ANNs trained on a database of drilling parameters to foresee potential problems during the drilling process.
Okpo et al. (2016)	Wellbore instability	The Niger Delta oil field was used as a case study for training ANNs to predict wellbore instability.
Ahmadi et al. (2016)	Prediction of mud density based on wellbore conditions	The quality of the drill bit was evaluated in real-time by an artificial neural network (wear). Non-linear, multi-input, multi-output drilling system was modelled using ANNs. Artificial neural networks were used to determine which data bit was "best". Bed heights in wellbores with varying degrees of inclination were predicted using ANNs. Utilized ANNs for bit bounce detection, which can be used to avert abnormal drilling conditions such as bit whirl and stick-slip. Drilling hydraulics optimization and prediction using artificial neural networks. To improve drill time estimations, ANNs were used to predict ROP. Improved ROP through the use of artificial neural networks for predicting and managing bit wear. Utilized ANNs trained on a database of drilling parameters to foresee potential problems during the drilling process. The Niger Delta oil field was used as a case study for training ANNs to predict wellbore instability. information gathered from published sources

(*Continued*)

TABLE 2.2 (CONTINUED)
Use of Neural Networks in Oil Exploration

Author(s)	Application	Notes
Elkatatny et al. (2016); Al-Azani et al. (2018) Abdelgawad et al. (2018)	Drilling fluid rheological properties	Drilling progress was monitored by artificial neural networks to determine the quality of the drill bit being used (wear). Non-linear, multi-input, multi-output drilling system was modelled using ANNs. Artificial neural networks were used to determine which data bit was "best". Bed heights in wellbores with varying degrees of inclination were predicted using ANNs. Utilized ANNs for bit bounce detection, which can be used to avert abnormal drilling conditions such as bit whirl and stick-slip. Drilling hydraulics optimization and prediction using artificial neural networks. To improve drill time estimations, ANNs were used to predict ROP. Improved ROP through the use of artificial neural networks for predicting and managing bit wear. Utilized ANNs trained on a database of drilling parameters to foresee potential problems during the drilling process. The Niger Delta oil field was used as a case study for training ANNs to predict wellbore instability. Information gathered from published sources. Drilling fluids' rheological properties were predicted using ANNs.
Vassallo et al. (2004)	Bit bounce detection	Utilized ANNs for bit bounce detection, which can be used to avert abnormal drilling conditions such as bit whirl and stick-slip.
Cristofaro et al. (2017)	Mud losses	To determine the most effective method of preventing mud losses, researchers used several different types of artificial intelligence.
Hoffmann et al. (2018)	Drilling reports sentence classifications	The use of ANNs allowed the creation of a methodology for the automatic organization of sentences from drilling reports into the following three tables: Events, Symptoms, and Actions. used information from 303 boreholes.
Li et al. (2018)	Lost circulation	Foretold the likelihood of lost circulation while drilling using ANNs.
Al-AbdulJabbar et al. (2018b)	Formation top prediction while drilling	Drilling operations assisted by artificial neural networks for predicting formation peaks.
Elzenary et al. (2018)	Equivalent circulation density (ECD) prediction	Predicted ECD while Drilling Using ANNs.

Source: Alkinani et al., 2019

TABLE 2.3
Implementations of ANN for Industrial Purposes

Author(s)	Application	Notes
Thomas and Pointe (1995)	Conductive fracture identification	Used ANNs to identify conductive fractures.
Denney (2000)	Fracturing restimulation candidates	Used ANNs to identify fracture restimulation candidates with case history from Red Oak field.
Faga and Oyeneyin (2000)	Gravel-pack design	In order to complete the gravel pack in the well in a timely manner, we used ANNs to get the grain size distribution data.
Al-Fattah and Startzman (2001)	Natural gas production prediction	Developed ANN model to forecast the United States gas production to the Year 2020.
Salehi et al. (2009)	Casing collapse due to production	Casing collapse issues due to reservoir compaction, poroelastic effects, and corrosion were predicted using ANNs. Also, an example from a substantial carbonate oil field in Iran was provided.
Adeyemi and Sulaimon (2012)	Wax formation prediction	Used ANNs to predict the wax formation.
Moradi et al. (2013)	Wax disappearance temperature	Used experimental and real data of wax precipitation.
Costa et al. (2014)	History matching	Used ANN for pattern recognition in a reservoir simulation model.
Yanfang and Salehi (2014)	Re-fracture candidate selection	Used real field data from the Zhongyuan oilfield.
Al-Naser et al. (2016)	Application of multiphase flow patterns	Used “Unified Model” to generate the data and used experimental data for testing.
Ghahfarokhi et al. (2018)	Prediction of gas production	Used ANNs to predict gas production in the Marcellus Shale.
Khan et al. (2018)	Oil rate prediction	Used ANNs to predict the optimum production rate.
Luo et al. (2018)	Production optimization	Used ANNs to optimize the production in the Bakken Shale.
Nande (2018)	Hydraulic fracturing	Used ANNs to minimize the error in predicting closure pressure for hydraulic fracturing analysis.
Nieto et al. (2018)	Completion optimization	Used ANNs to optimize the completion and to protect parent well in the Montney formation in British Columbia.
Pankaj (2018)	Well spacing and well stacking	Used ANNs to optimize well spacing and well tacking in the Permian Basin.
Sidaoui et al. (2018)	Carbonate acidizing	Used ANNs to predict the optimum injection rate of carbonate acidizing.
Tariq (2018)	Flowing bottom hole pressure prediction	Used ANNs to predict flowing bottom hole pressure.
Al-Dogail et al (2018); Basfar et al. (2018)	Inflow performance (IPR) prediction	ANNs were used to foretell how well gas fields and vertical oil wells would produce their inflows.

Source: Alkinani et al., 2019

ANNs in the petroleum industry. The following findings were reached after perusing a substantial body of literature on the topic of ANNs in the petroleum industry: the petroleum industry has access to massive amounts of historical data, which can be used to make accurate predictions about the future. Large uncertainties in the future make accurate prediction difficult. As a result, ANNs can be used for both long-term and short-term forecasting, with high accuracy, allowing decision-makers to plan ahead for the best possible outcomes. The analytical solution to some problems in petroleum engineering is challenging. As a result, ANNs can be a reliable tool for addressing these sorts of challenges as shown in Tables 2.1 to 2.4. The applications were categorized according to four distinct groups: reservoir, production, exploration, and drilling.

2.7 DIFFERENT TYPES OF SURVEYS

Geophysical information is crucial to the oil and gas industry in many ways. Picking the optimal drilling location could be as much art as science. Oil and gas are found by geophysicists by creating a detailed picture of what lies beneath the surface of the Earth, as most petroleum deposits are hidden there (Wongsinphaiboon, 2022).

Using geophysical techniques, such as the study of electrical currents, gravitational and magnetic anomalies, heat flow, geochemical relationships, and variations in Earth's density, oil can be located. Using data collected from numerous sources, it is possible to locate hydrocarbons buried deep within the Earth.

2.7.1 Seismic Surveys

The echoes that are reflected back to the surface after being bounced off of subterranean rock formations are recorded by sensors for later analysis. It is possible to learn a lot about the composition of rocks and the presence of gases and fluids within them by analyzing the amount of time the waves take to return. Ultrasound is used in a way that is analogous to this in the medical field. In order to create a seismic signal, onshore operations typically employ the use of specialized trucks transporting a heavy plate. A "seismic streamer", or series of cables with seismic sources and hydrophones attached, is towed by a specialized vessel during offshore operations. Compressed air is used in the seismic sources to generate sound. The sound waves that are reflected back to the hydrophones are recorded for later analysis (Wongsinphaiboon, 2022).

2.7.2 3D Seismic Survey

The most recent and cutting-edge technique is a three-dimensional seismic survey (3D). With the help of a computer, we can now see a full three-dimensional model of the subterranean landscape. It takes about $30,000 per mile to finish the process (Satterwhite, personal communication). However, drilling a well can cost millions of dollars, so it's often worthwhile to invest in extensive seismic surveys prior to breaking ground to identify promising areas. In addition, this strategy has been validated for its ability to foresee faults and overpressure, both of which can pose problems during drilling. Seismic surveys, whether on land or in water, can typically be conducted with minimal disruption to locals and no negative effects on the environment. Exploration geologists rely heavily on this method for finding fossil fuel deposits.

2.8 CLOUD TECHNOLOGIES

The NIST definition of cloud computing is "easy, on-demand network access to a shared pool of configurable computing resources". This model aims to make that possible (such as servers, storage devices, networks, applications, and services). Due to its low management cost and interaction service function, the cloud computing architecture can be quickly released and deployed (Xu and Zhao, 2013). Multiple well-established computing technologies have been combined to form the cloud. These include grid computing, distributed computing, network storage, virtualization, load balancing, and more (Vaquero and Rodero-Merino, 2008). One defining feature of cloud computing is the sharing of resources among a distributed group of computers (Yigit et al., 2014). Users of a cloud service have access to all available storage options, and the service's enterprise data centre allows for dynamic resource migration in response to fluctuating app demands. The concept of "everything-as-a-service", in which users have access to a complete system (including hardware, software, and the underlying network) in the form of services, is emerging as a natural extension of cloud computing (Lenk et al., 2009).

Xu has listed a few of the characteristics that can be used to describe cloud computing, which are just a few of the many that can be used. These include information technologies that are available at a low cost, the deployment of real-time dynamic resources, adaptability, standardization, and extremely large scale (Xu and Zhao, 2013). The oil and gas industry can reap benefits from computing in the cloud because it improves operational efficiency and maximizes the utilization of available resources. Second, the utilization of cloud computing can assist the information technology sector in reducing costs because it does not require any upfront costs, and it offers free downloads, installations, and unlimited service. Last but not least, the information management needs of the oil and gas sector can benefit greatly from the cloud's high availability and real-time data backup. For the fourth part, the demand for green IT is met by the cloud's virtual data centre's ability to lessen energy usage and carbon emissions.

2.8.1 Cloud Technologies in the Petroleum Industry

Lack of data integration, sharing of successes, and cohesive teams are just a few of the issues plaguing the petroleum industry. The petroleum industry is increasingly adopting cloud computing as a means to modernize its operations due to the many benefits it offers, such as the capacity to store and share massive amounts of data and the flexibility to integrate previously separate hardware and software components. Barriers between departments in the petroleum industry are broken down with the help of data integration technology, which then allows for more streamlined management of all data and information. By linking data repositories in different locations, cloud computing aims to provide a more streamlined data information service and sharing mechanism.

Cloud computing in the petroleum industry is typically divided into two categories: desktop cloud and research cloud. When academics use the cloud, they increase their efficiency in both discovery and innovation. Cloud computing enables the central office, the subsidiary companies, and even the most remote locations of the oil company to work together and visualize their data in real time. The petroleum industry is hoping that, by utilizing cloud services, it will be able to improve its existing supercomputing capabilities, process the massive amounts of seismic data produced by ultra-sensitive seismic sensors, and speed up the analysis of the images that are produced as a result of this processing. The majority of the discussions that have taken place regarding cloud computing in the oil and gas industry have centred on cloud-based desktop applications, management, and online software services that are utilized during the core of the production process. An illustration of how cloud computing is being put to use in the petroleum industry can be seen in Figure 2.4 (Zhifeng, 2019).

2.8.2 Cloud Storage Application of Petroleum Industry

With the advent of 3D seismic exploration and the refinement of high-precision 3D seismic acquisition techniques, the data volume associated with petroleum exploration has grown to be enormous. Seismic data is an example of a subset that can be classified as "big data". Storage devices with high performance and large capacities are necessary for the efficient management of seismic data, but a well-designed system is also essential. Monitoring seismic data is crucial for any Earth Science study. It can store data quickly, has a large storage capacity, can adjust to a wide range of conditions, is highly reliable, and can preserve information for a long time. The use of cloud computing for the archival of large datasets has become widespread in recent years. Businesses are able to back up their data and access it from anywhere with the help of cloud storage. When fully implemented, cloud storage will be able to consolidate many different types of storage media into a single, streamlined service with the help of cluster applications, grid computing, and a distributed file system. A large-scale distributed cloud storage system possesses numerous desirable characteristics, such as high concurrent access, a large storage capacity, high efficiency, high performance, and dependable service for a file-sharing storage platform. Currently, cloud storage security and cloud storage design are the primary foci of research. Researchers in all fields are increasingly turning to cloud services for mission-critical data storage as their popularity rises.

Oil exploration is a tedious and time-consuming process that heavily relies on a single computer. The storage architecture developed for the oil and gas industry is highly dependable, secure, and sustainable because it makes use of a distributed parallel cloud storage system. It can accommodate the high I/O reading and writing bandwidth, the high floating point computing performance, and the storage requirements of the oil and gas industry's massive seismic data. The four main parts of a

TABLE 2.4
Reservoir Applications of Artificial Neural Networks

Author(s)	Application	Notes
An and Moon (1993); Long et al. (2016)	Reservoir characterization	Used ANNs for reservoir characterization.
Yang and Kim (1996)	Rock properties	Used accelerated ANNs to find rock properties.
Denney (2001)	Reservoir monitoring	Used ANNs to find the optimum relationship between pressure, saturation, and seismic data. The model was applied to the Statfjord field.
Alcocer and Rodrigues (2001)	Fluid properties	Used ANNs to estimate fluid properties using nuclear magnetic resonance.
Denney (2003)	Well testing	Used ANNs to analyze pressure transient data from an anisotropic faulted reservoir.
Denney (2006)	Uncertainties in reservoir performance	Used ANNs in Monte Carlo simulations to generate the probability distribution of possible outcomes.
Elshafei and Hamada (2007)	porosity and water saturation	Used ANNs to predict formation porosity and water saturation from well logs.
Ayoub et al. (2007)	Viscosity correlations	Used ANNs to evaluate the below bubble point viscosity correlations.
Al-Bulushi et al (2007); Khan et al. (2018)	Water saturation prediction	Used ANNs to predict water saturation for sandstone reservoirs using conventional well logs.
Hegeman et al. (2009)	Downhole fluid analysis	Used ANNs to estimate gas/oil ratio (GOR) from real-time downhole fluid samples.
Zabihi et al. (2011)	Permeability damage prediction	Used experimental data of Berea sandstone cores.
Kohli and Arora (2014)	Permeability prediction	Used ANNs to predict permeability from well logs.
Ma and Gomez (2015)	Predictions of hydrocarbon resource	Used real field data with some statistical methods and ANN.
Bello et al. (2016)	Drilling system design and operation	Used multiple artificial intelligence techniques—including ANNs—to design drilling and operation systems.
Li et al. (2018)	Geomechanical characterization	Used ANNs to successfully synthesize compressional and shear travel time logs.
Dang et al. (2018)	EOR	Used ANNs to perform N-dimensional interpolation of relative permeability.
Rashidi et al (2018)	Elastic modulus	Used ANNs to correlate between static and dynamic modulus of limestone formations. Also, used an example from two formations Asmari and Sarvak in Iran.
Hadi and Nygaard (2018)	Shear wave estimation	Used ANNs to predict shear wave in carbonate reservoirs.
Rashidi and Asadi,(2018)	Pore pressure estimation	Used ANNs to predict formation pore pressure from drilling data.
Hamam and Ertekin (2018)	CO_2 injection	Used ANNs to develop a screening tool for CO_2 injection in naturally fractured reservoirs.
Hasan et al. (2018)	Temperature distribution	Used ANNs to predict thermal distribution in thermal EOR methods.

Source: Alkinani et al., 2019

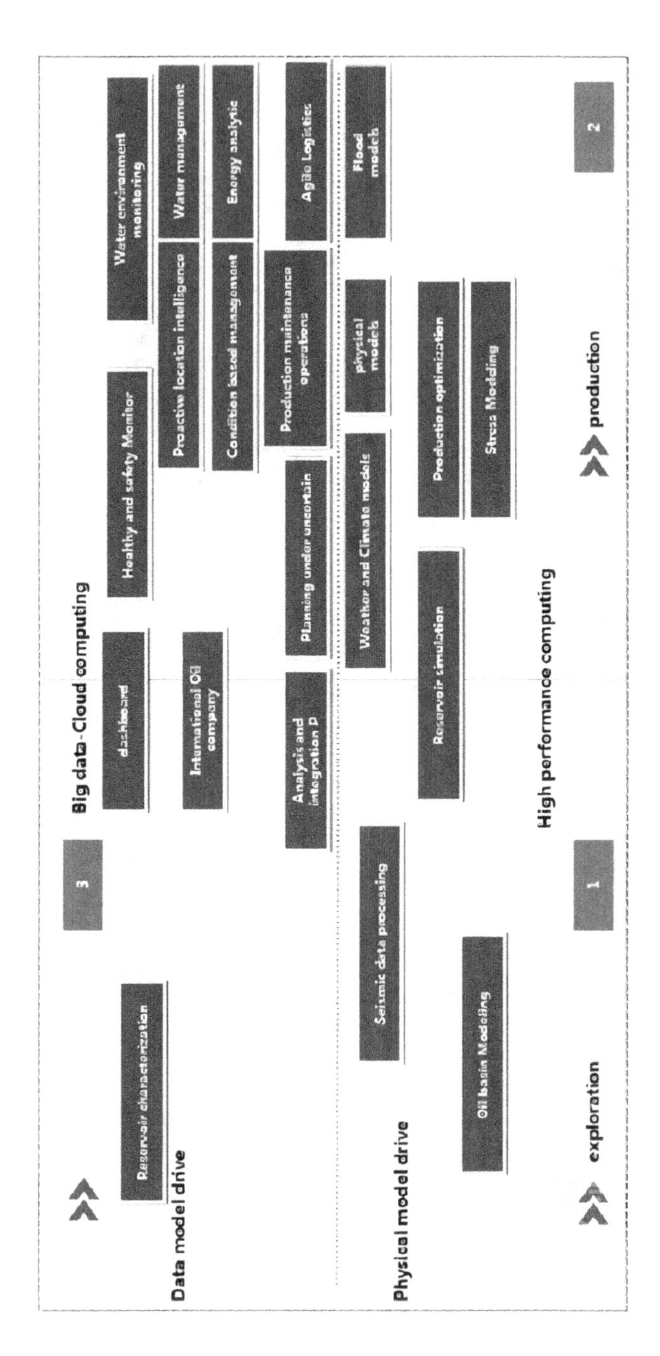

FIGURE 2.4 Cloud Computing of the petroleum industry (Zhifeng, 2019).

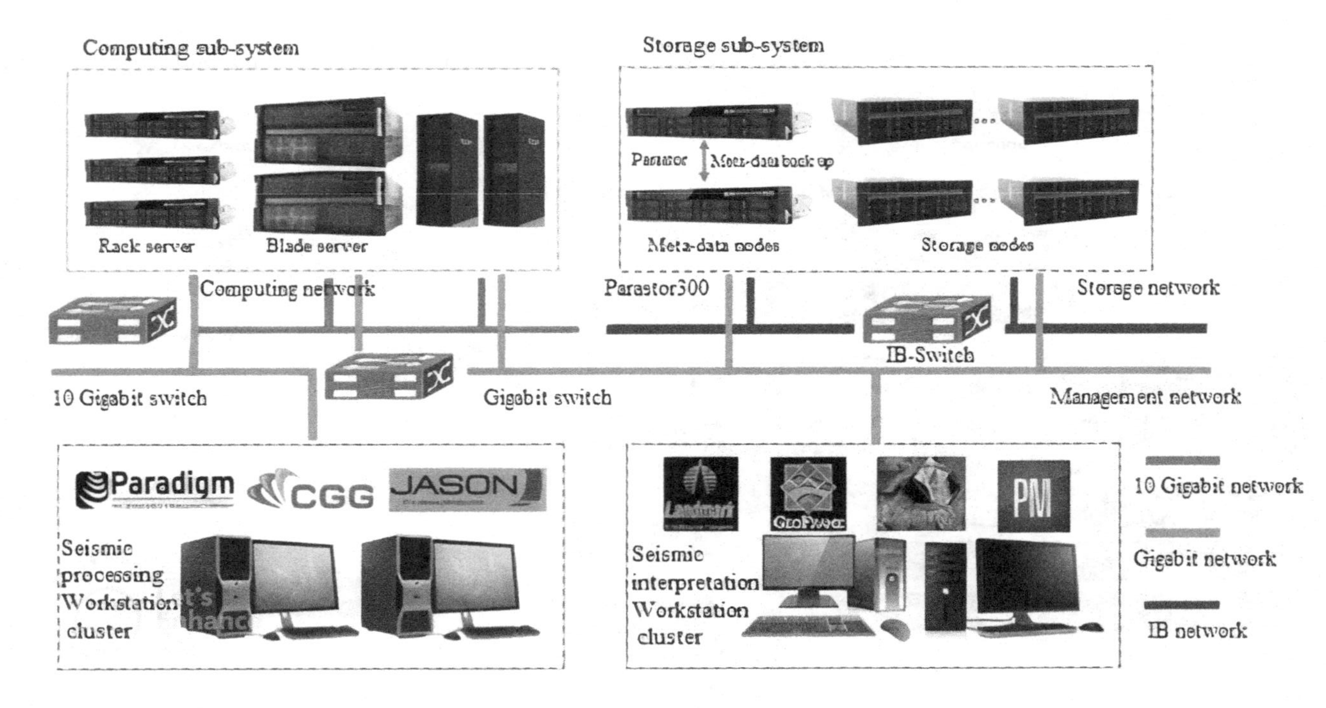

FIGURE 2.5 Seismic data storage: A distributed architecture (Zhifeng, 2019).

cloud storage architecture are the storage subsystem, the computing subsystem, the network subsystem, and the seismic data processing and interpretation subsystem. The storage subsystem includes a management node, a meta-data node, and a data node. The Portable Operating System Interface (POSIX) protocol is well suited for high-performance computing settings due to its fast data transfer rates and reliable cache. Processing seismic data is the primary task of the computing subsystem in oil production and exploration. The high floating point computing capability, scalability, and system memory bandwidth of blade servers and fat node servers make them common choices for seismic data processing. The whole design relies on a high-speed network of either 10GbE, FDR, or EDR to guarantee the rate of data transmission and to satisfy the needs of customers who demand such a connection. After seismic data processing is complete, the results are displayed on the workstation, at which point the underlying strata and faults are explained. This report serves as a resource for geological engineers to use in making drilling decisions. Seismic data is stored in a distributed architecture, as shown in Figure 2.5.

2.9 CONCLUSION

The term "fuelling" refers to the role that oil plays in the global economy, but oil is the primary non-renewable energy source. While the rising cost of international crude oil has certainly bolstered efforts to develop renewable energy sources, their share of global energy consumption remains negligible despite these efforts. Every facet of the global energy market and every region of the world are affected by the rapid development of new technologies. The extremely competitive and global petroleum industry encourages the global dissemination of technology. The advancement of technology contributes to the expansion and growth of the economy. Unavoidable, but notoriously difficult to predict, is the development of new technologies (Ike et al., 2013).

BIBLIOGRAPHY

Alkinani, H., Al-Hameedi, A. T., Dunn-Norman, S., Flori, R., Alsaba, M., & Amer, A. (2019, November). *Applications of Artificial Neural Networks in the Petroleum Industry: A Review.* https://doi.org/10.2118/195072-MS

Aminzadeh, F., & deGroot, P. (2005, January 1). *A Neural Networks Based Seismic Object Detection Technique.* Society of Exploration Geophysicists.

Bylov, G. V. (2013). Technical aspects of building GIS in petroleum geology. *Neftyanoe Khozyaystvo—Oil Industry*, 45–48.

Clifford, A., & Aminzadeh, F. (2011, January 1). *Gas Detection from Absorption Attributes and Amplitude Versus Offset with Artificial Neural Networks in Grand Bay Field.* Society of Exploration Geophysicists.

Noradila Rusli ESRI. (2007). *GIS Best Practices: GIS for Petroleum.* 1–32. ESRI.

Fogg, A. N. (2000, January 1). *Petro-Seismic Classification Using Neural Networks: UK Onshore.* Society of Exploration Geophysicists.

Guo, Y., Hansen, R.O., & Harthill, N. (1992, October). Artificial intelligence I neural networks in geophysics. Paper presented at the 1992 SEG Annual Meeting, New Orleans, Louisiana.

Hami-Eddine, K., Klein, P., & de Ribet, B. (2015, December 17). *Predicting Reliability of AVA Effects Using Neural Networks.* Society of Exploration Geophysicists.

Hansen, K. V. (1993, January 1). *Neural Networks for Primary Reflection Identification.* Society of Exploration Geophysicists.

He, Z. M., & Wu, X. C. (2003). Application of GIS in petroleum geology. *Jianghan Shiyou Xueyuan Xuebao/Journal of Jianghan Petroleum Institute, 25–27*(2).

Huang, K.-Y., Pissarenko, J.-D., Chen, K.-J., Lai, H.-L., & Don, A.-J. (2006, January 1). *Neural Network for Parameters Determination and Seismic Pattern Detection.* Society of Exploration Geophysicists.

Huang, K.-Y., & Yang, J.-R. (2015, December 17). *Seismic Velocity Picking Using Hopfield Neural Network.* Society of Exploration Geophysicists.

Ike, D. U., Anthony, A., Anthony, A., Abdulkareem, A., & Abdulkareem, A. (2013). Impact of ICT in oil and gas exploration: A case study. *International Journal of Computers & Technology, 10*(7). https://doi.org/10.24297/ijct.v10i7.3214

Karrenbach, M., Essenreiter, R., & Treitel, S. (2000, January 1). *Multiple Attenuation with Attribute-Based Neural Networks.* Society of Exploration Geophysicists.

Kononov, A., Gisolf, D., & Verschuur, E. (2007, January 1). *Application of Neural Networks to Travel-Times Computation.* Society of Exploration Geophysicists.

Lenk, A., Klems, M., Nimis, J., Tai, S., & Sandholm, T. (2009). What's inside the cloud? An architectural map of the cloud landscape. *Proceedings of the 2009 ICSE Workshop on Software Engineering Challenges of Cloud Computing, CLOUD* 2009. https://doi.org/10.1109/CLOUD.2009.5071529

Mohaghegh, S. (2000). Virtual-intelligence applications in petroleum engineering: Part I – artificial neural networks. *JPT, Journal of Petroleum Technology, 52*(9). https://doi.org/10.2118/58046-ms

Napatr Wongsinphaiboon. (2022, November 5). *Geophysical Surveys.* https://petgeo.weebly.com/geophysical-surveys-for-petroleum.html

Refunjol, X., Infante, L., & Bernaez, A. (2016, January 1). *Identifying Unconventional Potential Using Seismic Inversion and Neural Networks: An Eagle Ford Shale Study.* Society of Exploration Geophysicists.

Ross, C. (2017, October 23). *Improving Resolution and Clarity with Neural Networks.* Society of Exploration Geophysicists.

Russell, B., Ross, C., & Lines, L. (2002, January 1). *Neural Networks and AVO.* Society of Exploration Geophysicists.

Smith, A., & Hania, J. (2000). What is GIS? *New Zealand Journal of Geography*, 110–116.

Sun, Q., Castagna, J., & Liu, Z. (2000, January 1). *AVO Inversion By Artificial Neural Networks (ANN).* Society of Exploration Geophysicists.

Vaquero, L., & Rodero-Merino, L. (2008). A break in the clouds: Towards a cloud definition. ACM SIGCOMM Computer Communication Review , *39*(1), 50–55.

Verma, S., Roy, A., Perez, R., & Marfurt, K. J. (2012, November 4). *Mapping High Frackability and High TOC Zones in The Barnett Shale: Supervised Probabilistic Neural Network vs. Unsupervised Multi-Attribute Kohonen SOM.* Society of Exploration Geophysicists.

Xiangjun, Z., Youming, L., & Hong, L. (2000, January 1). *Hydrocarbon Prediction Using Dual Neural Network.* Society of Exploration Geophysicists.

Xu, W. P., & Zhao, H. (2013). Research of cloud computing information management mode for oil enterprise. *Applied Mechanics and Materials, 336–338.* https://doi.org/10.4028/www.scientific.net/AMM.336-338.2048

Yigit, M., Gungor, V. C., & Baktir, S. (2014). Cloud computing for smart grid applications. *Computer Networks, 70.* https://doi.org/10.1016/j.comnet.2014.06.007

Zhifeng, Y. (2019). Cloud computing and big data for oil and gas industry application in China. *Journal of Computers, 14*(4). https://doi.org/10.17706/jcp.14.4.268-282

3 Data Handling Techniques in the Petroleum Sector

Sonali Vyas, Shaurya Gupta, and Vinod Kumar Shukla

3.1 INTRODUCTION

The industry of oil and gas stands as a much-regulated production industry because of its intrinsic reasons pertaining to well-being, security, and ecological threat which are related by means of examination, penetrating, manufacturing, and dispensation dissemination accomplishments. These directing necessities, in accumulation to an emergent expertise gap because of the superannuation of knowledgeable personnel, in addition to low prices of oil for an extremely long period have compelled oil and gas corporations to be quite inventive and disrupting in order to augment efficiency and competence so as to moderate risk and minimize capital and functional expenses, and escalate incomes, in addition to progress with supervisory amenability. Considering the past period, there has been a speedy stride in terms of technical modernization and industry associated with technical implementation which have transformed from definitive commercial representations, to complete industrialized, monetary and socioeconomic settings. Contemporary progressions in terms of information and communication technologies, comprising cloud computing, extraordinary enactment workstations, great dimensional imagining competences, preservative industrialized manufacturing, self-directed robotic systems, have catalyzed digital implementation throughout businesses. With time, all these technologies have simplified cyber–physical incorporation involving data collection, analysis, and visualization so as to take more well-versed conclusions apart from serving as a platform for imitations or simulations in order to elevate procedures. The above discussed notion of cyber physical collaboration and imitation is usually called digital twin). Gartner Group, being a research organization, acknowledged digital twin as among the top planned technological developments (Mohr et al., 2018), (Panetta et al., 2017), (Barricelli et al., 2019), (Wu et al., 2021). Many business sectors, such as industrialized manufacturing (Semeraro et al., 2021), (Holler et al., 2016), (Qi et al., 2018), (Kritzinger et al., 2018) motorization (Zhang et al., 2017), (Tao et al., 2018), well-being care (Feng et al., 2018), aeronautics, and

DOI: 10.1201/9781003357872-3

global exploration, have successfully deployed digital twin. Oil and gas commerce have influenced digital expertise in order for commercial and operational models to generate new income and value generating prospects. Acceptance of digital expertise and transforming to a digital commerce is called digitalization. With implementation of emergent digital expertise, oil and gas businesses are focusing on digital twin technology. Generally, its implication in business is carried out through a bottom-up method, wherein the technological expertise is unsymmetrically executed. In order to exploit the full potential of digital twin and associated technical implementations, a complete considerate understanding in terms of digital twin technology is required apart from exploration accomplishments with the prospects and encounters allied with implementation of digital twin in the oil and gas industry.

3.2 BIGDATA IN O&G

Considering the past few years, energy corporations have capitalized on software for the purpose of seismic data dispensation and imagining, which in turn helps them in understanding what all elements are like beneath Earth's surface, as well as the numerous ways of extracting them. This allows for an acceptance of BigData expertise in terms of oil and gas. However, the associated facilities and systems accompanying BigData disposition advances many cybersecurity and data confidentiality concerns. Figure 3.1 discusses the application of BigData with regard to smart oilfields (Settemsdal et al., 2019), (He et al., 2016).

Apart from that, lack of checking and monitoring tools in contemporary systems marks it to be quite defenceless in case of cyber-attacks (Mittal et al., 2017). BigData denotes capacious datasets, usually of petabytes or exabytes (Mohammadpoor et al., 2020). Data specialists are responsible for excerpting significant understandings from raw data which are quite critical for the purpose of results for varied business applications. Equinor, a petroleum refining organization, primarily unveiled all subsurface and operational information from a field on an inland ledge in order to sustain

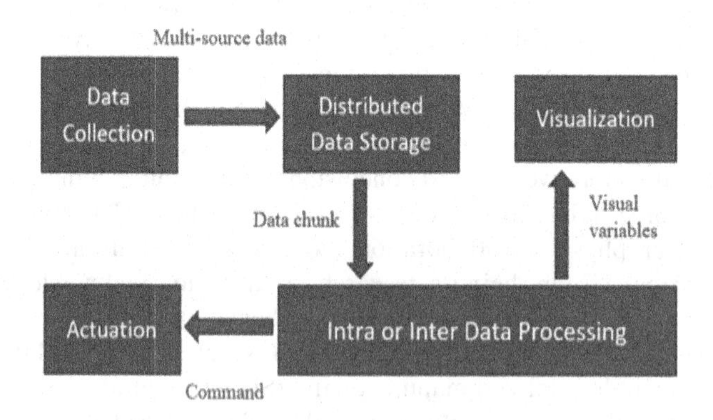

FIGURE 3.1 Application of BigData in smart oilfields.

education improvement, as well as novel resolutions for imminent energy abstraction. The datasets include data of geophysical clarifications, such as the following:

- Geo science oil well documentation.
- Seismic well record.
- Production pool models.
- Real-time boring information.

Numerous trainings have made an honest determination in terms of defining features of BigData, comprising techniques for accumulating information and scrutinizing in terms of excerpting valued info. IBM emphasizes the following three V's:

- Volume.
- Velocity.
- Variety for their applications.

Countless educations have designated the assistances of BigData deployment in oil and gas businesses. Table 3.1 shows prominent BigData applications in the oil and gas supply chain.

Considering the case of upstream oil and gas, current developments of seismic procedures, station counting, front observing geophones, logging during the process of penetrating the Earth's surface have enhanced data amount considerably. BigData analytics provide a very effective resolution in managing and analyzing these types of data. Xia et al., 2017, discussed replacing conservative tools in accordance with BigData procedures and the Hadoop platform to analyse enormous seismic datasets, recognize precarious environmental topographies, and describe reservoirs and geological subjects. Considering the case of midstream oil and gas, BigData lessens the cost and carbon dioxide emanations in cases of shipping conveyance, and supports the observing and upkeep of channel pipeline structures. The existing data can be compared with data sources like past data, upkeep reports, operative data, and analytic structure, which is able to sense and confine any kind of leak in a pipeline structure

TABLE 3.1
BigData Applications in the Oil and Gas Supply Chain

UPSTREAM	Investigation and exploration
	Oil production
	Reservoir Engineering
MIDSTREAM	Pipeline transference
	Ship conveyance
DOWNSTREAM	Refining
	Health and safety supervisory
	Sales and trade

(Joshi et al., 2018), (Layouni et al., 2017), (Liu et al., 2013). This same approach is also functional in terms of gear conservation in varied sectors. Data investigation delivers performance in terms of gear functioning and the prediction of perilous procedures. The progressive utilities are valuable for upkeep preparation, risk extenuation, and safety administration through oil and gas procedures. Other than that, it augments the large-scale supervision of oil and gas assets (Alobaidi et al., 2015), which includes exploiting functioning competence, comprising energy and resource ingestion; apart from that, it upsurges unit consistency and advances the economic enactment of refinery. A good number of scholars have exploited the methodical and analytical competence of BigData in oil and gas sales and trading (Brelsford et al., 2018), (Chen et al., 2017), (Bekiroglu et al., 2018), (Desai et al., 2021).

3.3 BIGDATA ADMINISTRATION IN THE OIL AND GAS SECTOR

Currently, because of digitization, many oil and gas corporations have by far augmented competences in terms of observing and recording, in addition to analyzing information quite proficiently with the help of progressive technological expertise (Attia et al., 2020). Whereas seismic data interpretation showcases the consideration of data in terms of high-performance mainframes and progressive imagining procedures. It involves data attainment and data dispensation processes which are discussed below.

3.3.1 Data Attainment

Attainment of seismic information comprises the use of an enormous amount of seismic sensors which are called geophones: ground gesticulation sensors responsible for converting ground sensations into energies by apprehending replicated waves which range from 10–100 Hz and are sent by a shuddering source. Sensors are regularly situated over enormous zones with the help of seismic restraints limiting suppleness and upsurging disposition charge (Song et al., 2019). In order to discuss such encounters, the use of wireless geophone sensor networks (Roden et al., 2016) along with subsurface cameras (Nguyen et al., 2020) is projected. Apart from that it uses reconfigurable antenna, wireless nodes, and entryways for seismic information so as to overcome the challenges accompanying wired seismic cables. Though these kinds of approaches are very promising, there exist numerous open concerns accompanying the usage of geophysical sensor systems, including noise issues, power ingestion, and short-range message exchange.

3.3.2 Data Processing

As the enormous data is produced from varied seismic aspects, different artificial intelligence and machine learning procedures are engaged for the purpose of improving estimation evaluations (Olneva et al., 2018) as analysis and inferring data with predictable approaches like seismic amplitude readings can be quite puzzling and imprecise. The authors in Souza et al., 2020, have showcased use of BigData analytics and use of artificial intelligence in cases of seismic data examination. Usage of unsupervised BigData and machine learning investigative procedures for seismic information breakdown deliver a fair understanding of geologic configurations. In Zhang et al., 2020, the authors actively used unsupervised machine learning constituent; i.e., principal component analysis for collection in terms of seismic characteristics and self-organizing maps for the purpose of cataloguing and explanation of seismic information.

It inculcates the following stages:

- Documentation of environmental concerns.
- PCA applications for the purpose of selection in terms of seismic characteristics.
- SOM ML contrivance used in classifying regular data clusters.
- 2D colour map used in identifying bunches in categorizing normal outlines.

Usage of PCA and SOM methodologies contributes enhanced threat valuation and clarification validated by geoscientists. Studies carried out by authors in Li et al., 2022, established the benefits of BigData methodical and machine learning procedures for the purpose of unearthing novel exploration; apart from that, researchers used machine learning procedures in classifying distinct substances in seismic and biological configurations within some provincial databases.

3.4 CONTEMPORARY FRAMEWORKS IN THE OIL AND GAS SECTOR

With time, many structures or frameworks have come to be developed for digital twin. The maximum acknowledged framework comprises of following constituents which are being showcased in Figure 3.2 (Ismail et al., 2022):

- Physical cosmos.
- Virtual cosmos.
- Connections in between spaces.

Physical cosmos encompasses physical quality, sensors and actuators, whereas virtual cosmos embraces multiscale, probabilistic simulation prototypes, and amassing

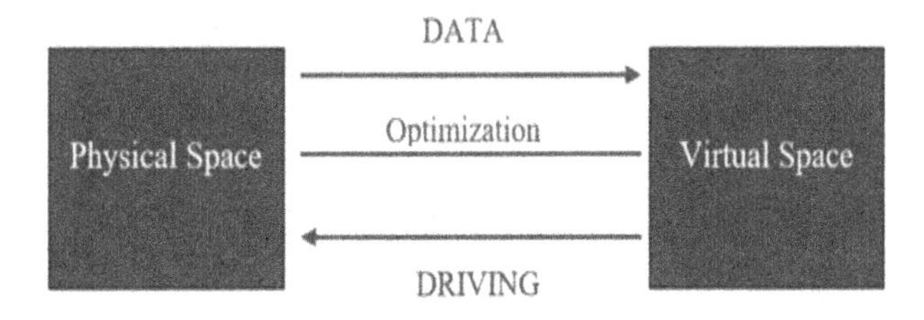

FIGURE 3.2 Contemporary frameworks in the oil and gas sector.

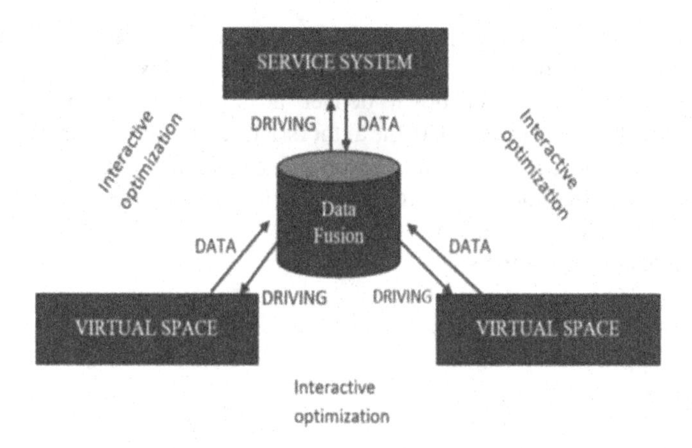

FIGURE 3.3 Digital twin data fusion components and services.

and examining information in addition to performing recreations so as to regulate optimum device restrictions plus circumstances for physical strength. The interconnection in between physical and virtual cosmoses guarantees unified actuation command interchange in between two spaces. Lately, this model was protracted to five constituent agendas including digital twin data fusion, and service methods; communication amongst them is discussed in Figure 3.3 (Grieves et al., 2020). Virtual cosmos acts like digital emulation for great reliability simulation of corporeal equivalent. Service organization encompasses other innovative software gears like imagining amenities, product superiority facilities, investigative facilities, model standardization amenities. The fusion model functions as a connection between physical and virtual cosmos which is responsible for collecting sensor data from physical cosmos, simulation data from virtual cosmos and service systems. Assorted data is attached and investigated by security procedures and encryption in order to defend digital twin and sensor information in contradiction of cyber-attacks.

The iterative phases of digital twin centred procedures involve creating, collaborating, collectively, analysis. Physical plants are equipped with sensors in order to generate electrical indications which characterize operative and conservational situations. Therefore, the real data is amassed with other current information, such as bill of supplies, design stipulations, engineering data and drawings, as well as event logs (Cheng et al., 2020). Progressive analytics and imagining gears, like machine learning, BigData analytics, virtual reality, and augmented reality, are engaged in order to investigate composed information apart from envisaging outcomes.

3.5 BIGDATA PROSPECTS AND CHALLENGES IN THE OIL AND GAS SECTOR

There are varied opportunities and challenges while discussing the role of digital twin in the oil and gas sector. Some of the prospects are discussed below.

3.5.1 Asset Enactment Controlling

Digital twin is responsible for obtaining information from oil and gas possessions and analysis of information in real-time or near real-time situations. These perceptions include production proportions, method blockages, operative circumstances, breakdown, and regulatory restrictions in order to augment manufacturing. Additionally, digital twin delivers a solitary interface for the purpose of imagining possibilities and significant enactment pointers of assets. Operatives have the facility of utilizing varied comprehensions and signs in order to augment fabrication for the purpose of maintenance and substitutions, keeping in mind "what-if" circumstances so as to increase production, while dropping downtime.

3.5.2 Asset Risk Valuation

Digital twin employs deep-learning and artificial intelligence procedures in order to sense and correct asset breakdowns. Services of individual physical assets connected to digital twin are examined uninterruptedly in classifying probable fiascos and evading all kinds of accidents. Whenever an engineering team is developing any kind of operative practices for augmenting manufacture, it can be checked with digital twin to scrutinize gears fitted in capacity whether it pertains to any kind of peril towards any human.

3.5.3 Virtual Operational Drill

Varied day to day operations in any oil and gas sector rely completely on workers' training and understanding in that facility. Worldwide, these companies are fronting challenges of these kinds (Wanasinghe et al., 2020) wherein a greater number of experienced workers are going to retire in the near future, thereby forming a scarcity of skilled workers throughout the whole industry. Therefore, operative drill and teaching courses are very much desirable in terms of orientation of fresh workers so that the experienced workforce can impart knowledge to newer workers. Digital twin, in collaboration with virtual and augmented reality, provides a good simulated platform in terms of training new employees to be quite operative in terms of navigation amongst oil and gas amenities, keeping in view the monitoring, inspecting varied devices, and efficient interaction amongst ongoing operations.

3.5.4 Disaster Reaction Training

Regular safety training, like fire manoeuvres and disaster emigration exercise, are compulsory in the oil and gas industry; being familiar with drill exercises, gear, and when equipment should be shut down are also very important (Parshall et al., 2017). Furthermore, alternative emigration exercises on offshore amenities are extremely expensive and may attract needless dangers and threats to workers working on board.

Some of the challenges are discussed below:

3.5.5 Lack of Standardization

Data or information stands as a mainstay in terms of digital twin, though it might not follow a mutual custom as it may be unstructured, semi-structured, or structured.

Information incorporation platforms from diverse retailers will also trail dissimilar criteria and approaches in data presentations. Furthermore, prevailing information is characteristically not associated to common databanks being stored in dissimilar positions. These aspects generally make it thought-provoking while considering the incorporation of all present and real-time information into a solo data systematic component.

3.5.6 Data Ownership and Sharing

The information produced amongst a digital twin may be used for upcoming expansions, comprising some of the additional possessions for the purpose of enlargement of progressive algorithms dealing in data science. With time, data will be playing a critical role as more and more industries are becoming digitalized (Bridle et al., 2018). Technical titans like GE as well as Siemens are evolving with the concept of digital twin for oil and gas business (Loi et al., 2017), whilst the oil and gas providers and operators will be making use of the produce. Simplicity along with precision in accordance to data proprietorship is acute, as is protecting intellectual property rights, which need to be considered in terms of enabling protected admittance, in apportioning accountabilities and describing the level of access for shareholders while data sharing (Priyanka et al., 2022).

3.5.7 Functionality

While considering the case of digital twin, it characteristically amasses a massive quantity of raw information from oil and gas resources, which in turn aim to engender intuitions regarding assets. Oil and gas workers must have admittance to data as and when required without being abstracted by any sort of other data which has been generated by digital twin (Lu et al., 2019). Consequently, the engineers of digital twin should comprehend the need for oil and gas operatives apart from their necessities preceding the design process.

3.6 RAMI Dossier in the Case of the Oil and Gas Industry

Generally, handling in case of upstream oil and gas business is classified in two types:

- Offshore.
- Onshore.

The above two zones are geologically alienated in terms of headquarter flexibility and interconnection contributing to the chief aspect of a positive and successful procedure (Alvarez et al., 2018).

3.6.1 Requirement of Flexibility in the Upstream Oil and Gas Business

Wireless technology stands as a noteworthy achievement in the case of Industry 4.0 (Hall et al., 2022) as it has empowered interconnection amongst varied transportable accomplishments. Broader analysis, sophisticated flexibility, and movement are some of the characteristics of the various tasks in the upstream oil and gas business. Most upstream oil and gas operations in many countries are positioned offshore. It is the responsibility of a telecommunication supplier to provide a system that is precisely assembled and positioned to ensure connectivity amongst varied fields, manufacturing, and workplace areas. Temporarily, the most imperative component in terms of Industry 4.0 relies on the persistent flow of information amongst the constituents of commerce (Hermann et al., 2019). The charting of an upstream value chain of oil and gas in RAMI 4.0 is conducted so as to guide development complying with varied verticals of Industry 4.0.

3.6.2 Identification of Mobility

With the help of associating a value chain in the upstream oil and gas industry with RAMI 4.0, some alterations in terms of value chain implementation are discussed in Figure 3.4.

Transportable computing in terms of security in agreement with risk administration supervision is important for business. Evaluation is centred considering the hierarchy of RAMI 4.0 (Aripin et al., 2020), and the generation of diverse consequences throughout evaluation.

3.6.3 Readiness Assessment Framework

The basic conditions for transformation in this specific industry involves security and mobility, wherein the security aspect is one of the most critical considerations for the upstream oil and gas sector because of great business peril. Therefore, the

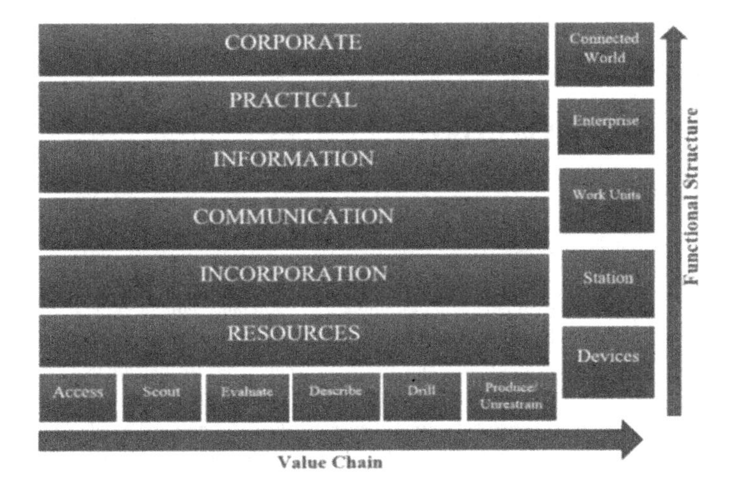

FIGURE 3.4 Upstream oil and gas industry with RAMI 4.0.

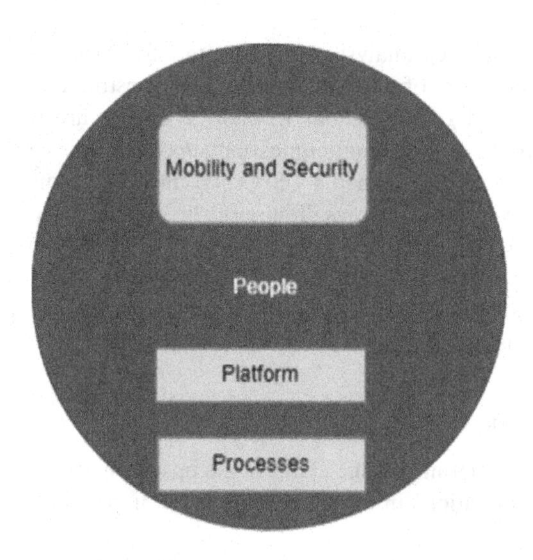

FIGURE 3.5 Readiness assessment framework.

readiness assessment framework focuses completely on the mobility context which is showcased in Figure 3.5.

Safety and protection are an important vertical of covering mobility. The operation of mobile computing trails all the safety and security criteria in agreement with risk administration assessment amongst the industry. These kinds of evaluations, when executed by RAMI 4.0, are managed and administered by a separate hierarchy and thereby generate various kinds of results (Aripin et al., 2020). Every evaluation at every hierarchy generates some kind of assessment score which is considered based on a balance of cost and probable worth. Table 3.2 deliberates an assessment outline for devices.

TABLE 3.2
Assessment Framework of Devices

Characteristics	Query themes	Objective
Individuals	Expertise knowledge in deploying, upholding novel technologies	Assessing the existing conditions and expectations in terms of acceptance and adaption of new technology
Individuals	Upskilling with the help of training or recruitment of trained workforce	Obtaining suitable skill set along with additional cost multipliers being incurred
Platform	Comparison of existing and contemporary digital technologies	Obtaining the magnitude of acceptance of novel technologies
Platform	Tools and operational services being used and whether to be purchased or hired on lease	Time descriptions which are directly or indirectly aligned with tools and internal resources

TABLE 3.3
Assessment Framework of Station

Characteristics	Query themes	Objective
Process	Taking consideration of green and brown fields	Contemporary number of oil and gas field administered apart from whether new technology should be adopted or not
Process	Operational services of a station is using a third party resource or some kind of internal resource	Whether the replacement of technology is full or partial and will it be a contractual replacement or a technological purchase
Platform	Work approval is digital or manual	Known or existing process can be an obstacle or some kind of anticipation
Platform	Availability of coverage of communications networks	Contemporary availability and need for enhancement of digital communication network

Table 3.3 discusses the assessment framework of varied stations dealing in the oil and gas industry.

The indispensable commercial prototype operational in the prevailing oil and gas industry encompasses (Barbosa et al., 2022) generating investments in the following verticals:

- Terrestrial privileges.
- Geological information.
- Drilling services.
- Human technical proficiency.

The model generally includes the following risk factors (Bjerga et al., 2022):

- **Dry-hole risk:** Capitalizing in investing dollars with no subsequent income from oil or gas because the penetrated geologic formation is a dry hole.
- **Drilling possibility:** Generally a high drilling expense may hamper a project's cost-effectiveness. Though establishments put in their best efforts in their perfect valuation, surprising ecological or automated complications may generate a substantial difference in definite budgets.
- **Production peril:** While drilling for natural oil and gas reserves, there stands a great possibility to estimate the sale and recoverability of hydrocarbon reserves over a period of time.
- **Expense peril:** The product expenses can have a drastic fluctuation which is very unexpected through any kind of momentous political happenings; for instance, conflict amongst nation-states, overproduction by the OPEC lobby, disruptions in sources such as a huge processing plant's equipment, labour slowdowns, or party-political revolutions.

- **Governmental peril:** Momentous quantities of the ecosphere's hydrocarbon assets are organized by countries ruled or governed by unbalanced governments. Establishments which participate in the developments of these nations must consider that administrations and privileged people amongst whom the contracts have been signed may or may not be in power at the time that sharing the earned profits comes about (which will need to be shared as per the contractual agreement). Considering many renowned circumstances, commercial reserves in terms of possessions, foliage, in addition to equipment, are basically municipalized by native administrations, which in turn may leave organizations deprived of any kind of revenue.

Oil and gas reserves are usually quite capital exhaustive by nature. Commercial elements and corporations ensure their existence based on the capability of proper interpretation of all such risks in accordance with their investment resources, which in turn confirms values to stakeholders (Asal et al., 2022). In order to emphasize the prominence of risk management in industry, numerous outsized oil corporations assign high-level business boards full of experts for reviewing and recommending risk assessments.

3.7 MAJOR RISKS TO COMPANIES

While considering the various types of risks (Oliva et al., 2022) for any oil and natural gas organization, it is always important to consider the following conditions:

- The importance level of an individual data source which is at a risk level.
- The number of times a certain risk is being cited in sources which are considered for a case study.
- The feasibility in terms of risk modelling.

Table 3.4 discusses the prioritized list of risks.

There are many other kinds of risks other than the risks (Samimi et al., 2020) discussed in the above table:

- Insurance risk.
- Foreign exchange risk.
- Credit risk.
- Legal and regulatory requirement risk.
- Joint venture risk.

Generally, a stochastic model is used in any of the modelling approaches because it uses a statistical distribution in possible representation regarding varied verticals which are catering to the possible consequences for any organization. Apart from that, it enables the analyst to see what outcomes may have a higher occurrence and enables the ability to model the relationships amongst certain types of risks. The modelling processes involves the following steps:

TABLE 3.4
Prioritized Risk List

Category	Risk	Description	Modelled risk
Economic	Product price	Volatility and dip in prices of oil and natural gas finally leading to poor operational results	YES
Environmental	Natural calamity	Any kind of likely catastrophe which is all natural and may affect the contemporary functioning of the system	YES
Operational	Industrialized calamity	Main mishap or oil tumble causing loss of life	YES
Political	Radical unpredictability	Resource disturbance because of conflict, civil combat, or terrorist attack	YES
Resource	Reserve obtainability	Change in material estimates because of some sort of potential development	YES

- Modelling risks involved.
- Calculation of degree of risks involved.
- Impact of risks in terms of an organization's monetary pointers.

Generally, the financial indicators are as follows:

- Cash and its equivalents.
- Capital return.
- Prices of shares.
- Profit calculation after tax.

3.8 RISK ANALYSIS /ASSESSMENT FOR ORGANIZATION

Considering a scenario of any organization the risk analysis of the below mentioned risks can be deliberated in detail (Karami et al., 2020; Jagoda et al., 2019).

- **Commodity price:** Considering the revenue of any organization which directly depends on the annual average level of selling oil and natural gas, and which are among the most vigorously operated merchandises, the prices fluctuate depending on many factors. The prices of all the commodities may fluctuate from time to time depending on various conditions and their corresponding hedging practices.
- **Natural disaster:** If an organization is working or functioning in some storm- or earthquake-prone area or region, then each and every event being conducted successfully at its level would be quite different compared to its

counterparts functioning in non-earthquake zones. This also has a large impact on insurance.

- **Industrial accident:** Risk management and safety procedures should be taken care off while industrial operations are being carried out as some kind of catastrophic risk might intimidate or threaten the feasibility of an organization.
- **Political instability:** The political conditions and scenarios in any nation should be monitored because they directly or indirectly affect the workings of an organization.
- **Resource availability:** It is quite possible for all sorts of new projects to operate with much lower revenue volumes than expected because of resource unavailability or resource outage.

The above mentioned points are the major topics which are given preference for any risk analysis of any organization and is considered and evaluated before venturing out to any projects. In the contemporary risk scenarios, these above points are the major issues which if not considered and properly managed can lead to failure of the project.

BIBLIOGRAPHY

Alobaidi, W. M., Alkuam, E. A., Al-Rizzo, H. M., & Sandgren, E. (2015). Applications of ultrasonic techniques in oil and gas pipeline industries: A review. *American Journal of Operations Research*, 5(4), 274.

Álvarez, E., Bravo, M., Jiménez, B., Mourão, A., & Schultes, R. (2018). The oil and gas value chain: A focus on oil refining. Cuadernos Orkestra, 2018/46, ISSN 2340-7638 (1–92).

Aripin, A. I., Abimanyu, A., Prabowo, F. S., Priandika, B., Sulivan, B., & Zahra, A. (2020, August). Mobile cloud computing readiness assessment framework in upstream oil and gas using RAMI 4.0. In *2020 International Conference on Information Management and Technology (ICIMTech)* (pp. 130–135). IEEE.

Asal, M., Mahar, K. M., Mahmoud Beshr, D., & Fawzy, S. F. (2022). Industry 4.0 technologies adoption: The impact of digital leadership and organizational learning on value-based digital transformation an applied study on Egypt's oil and gas sector. *Webology*, 19(3), 1284–1297.

Attia, H., Gaya, S., Alamoudi, A., Alshehri, F. M., Al-Suhaimi, A., Alsulaim, N. ... Al-Dirini, F. (2020). Wireless geophone sensing system for real-time seismic data acquisition. *IEEE Access*, 8, 81116–81128.

Barbosa, W. S., Gouvea, F. C., Martins, A. R. F., Belmonte, S. L., & FF, R. (2022). Development of spare-parts process chain in oil & gas industry using Industry 4.0 concepts. *International Journal of Engineering and Technology*, 14–16 (2).

Barricelli, B. R., Casiraghi, E., & Fogli, D. (2019). A survey on digital twin: Definitions, characteristics, applications, and design implications. *IEEE Access*, 7, 167653–167671.

Bekiroglu, K., Duru, O., Gulay, E., Su, R., & Lagoa, C. (2018). Predictive analytics of crude oil prices by utilizing the intelligent model search engine. *Applied Energy*, 228, 2387–2397.

Bjerga, T., & Aven, T. (2016). Some perspectives on risk management: A security case study from the oil and gas industry. *Proceedings of the Institution of Mechanical Engineers, Part O: Journal of Risk and Reliability*, 230(5), 512–520.

Brelsford, R. (2018). Repsol launches big data AI project at Tarragona refinery. *Oil & Gas Journal*, 116. https://www.ogj.com/refining-processing/refining/operations/article/17296578/repsol-launches-big-data-ai-project-at-tarragona-refinery

Bridle, J. (2018). Opinion: Data isn't the new oil—It's the new nuclear power. Ideas.Ted.Com.

Chen, Y., He, K., & Tso, G. K. (2017). Forecasting crude oil prices: A deep learning based model. *Procedia Computer Science*, 122, 300–307.

Cheng, J., Zhang, H., Tao, F., & Juang, C. F. (2020). DT-II: Digital twin enhanced Industrial Internet reference framework towards smart manufacturing. *Robotics and Computer-Integrated Manufacturing*, 62, 101881.

Desai, J. N., Pandian, S., & Vij, R. K. (2021). Big data analytics in upstream oil and gas industries for sustainable exploration and development: A review. *Environmental Technology & Innovation*, 21, 101186.

Feng, Y., Chen, X., & Zhao, J. (2018). Create the individualized digital twin for non-invasive precise pulmonary healthcare. *Significances Bioengineering & Biosciences*, 1(2).

Grieves, M. (2014). Digital twin: Manufacturing excellence through virtual factory replication. *White Paper*, 1(2014), 1–7.

Hall, R., Schumacher, S., & Bildstein, A. (2022). Systematic analysis of industrie 4.0 design principles. *Procedia CIRP*, 107, 440–445.

He, Y., Yu, F. R., Zhao, N., Yin, H., Yao, H., & Qiu, R. C. (2016). Big data analytics in mobile cellular networks. *IEEE Access*, 4, 1985–1996.

Hermann, M., Bücker, I., & Otto, B. (2019). Industrie 4.0 process transformation: Findings from a case study in automotive logistics. *Journal of Manufacturing Technology Management* (935–953). DOI:10.1108/JMTM-08-2018-0274

Holler, M., Uebernickel, F., & Brenner, W. (2016, October). Digital twin concepts in manufacturing industries-a literature review and avenues for further research. In *Proceedings of the 18th International Conference on Industrial Engineering (IJIE)* (pp. 1–9). Seoul: Korean Institute of Industrial Engineers.

Ismail, A., Ewida, H. F., Nazeri, S., Al-Ibiary, M. G., & Zollo, A. (2022). Gas channels and chimneys prediction using artificial neural networks and multi-seismic attributes, offshore West Nile Delta, Egypt. *Journal of Petroleum Science and Engineering*, 208, 109349.

Jagoda, K., & Wojcik, P. (2019). Implementation of risk management and corporate sustainability in the Canadian oil and gas industry: An evolutionary perspective. *Accounting Research Journal*.32(3), 381–398. https://doi.org/10.1108/ARJ-05-2016-0053

Joshi, P., Thapliyal, R., Chittambakkam, A. A., Ghosh, R., Bhowmick, S., & Khan, S. N. (2018, March). Big data analytics for micro-seismic monitoring. In *Offshore Technology Conference Asia*. OnePetro.

Karami, M., Samimi, A., & Jafari, M. (2020). Necessity to study of risk management in oil and gas industries (case study: oil projects). *Progress in Chemical and Biochemical Research*, 3(3), 239–243.

Kritzinger, W., Karner, M., Traar, G., Henjes, J., & Sihn, W. (2018). Digital twin in manufacturing: A categorical literature review and classification. *IFAC-PapersOnLine*, 51(11), 1016–1022.

Layouni, M., Hamdi, M. S., & Tahar, S. (2017). Detection and sizing of metal-loss defects in oil and gas pipelines using pattern-adapted wavelets and machine learning. *Applied Soft Computing*, 52, 247–261.

Li, S., Liu, N., Li, F., Gao, J., & Ding, J. (2022). Automatic fault delineation in 3-D seismic images with deep learning: Data augmentation or ensemble learning? *IEEE Transactions on Geoscience and Remote Sensing*, 60, 1–14.

Liu, Z., & Kleiner, Y. (2013). State of the art review of inspection technologies for condition assessment of water pipes. *Measurement*, 46(1), 1–15.

Loi, M., & Dehaye, P. O. (2017). If data is the new oil, when is the extraction of value from data unjust? *Filosofia e Questioni Pubbliche*, 7(2), 137–178.

Lu, H., Guo, L., Azimi, M., & Huang, K. (2019). Oil and gas 4.0 era: A systematic review and outlook. *Computers in Industry*, 111, 68–90.

Mayani, M. G., Svendsen, M., & Oedegaard, S. I. (2018, April). Drilling digital twin success stories the last 10 years. In *SPE Norway One Day Seminar*. OnePetro.

Mittal, A., Slaughter, A., & Zonneveld, P. (2017). *Protecting the Connected Barrels. Cyber Security for Upstream Oil and Gas*. Deloitte Insights.

Mohammadpoor, M., & Torabi, F. (2020). Big data analytics in oil and gas industry: An emerging trend. *Petroleum*, 6(4), 321–328.

Mohr, J. P. (2018). Digital twins for the oil and gas industry. Hashplay, San Francisco, CA, USA, Tech. Rep.

Nadhan, D., Mayani, M. G., & Rommetveit, R. (2018, August). Drilling with digital twins. In *IADC/SPE Asia Pacific Drilling Technology Conference and Exhibition*. OnePetro.

Nguyen, T., Gosine, R. G., & Warrian, P. (2020). A systematic review of big data analytics for oil and gas Industry 4.0. *IEEE Access*, 8, 61183–61201.

Oliva, F. L., Teberga, P. M. F., Testi, L. I. O., Kotabe, M., Del Giudice, M., Kelle, P., & Cunha, M. P. (2022). Risks and critical success factors in the internationalization of born global startups of Industry 4.0: A social, environmental, economic, and institutional analysis. *Technological Forecasting and Social Change*, 175, 121346.

Olneva, T., Kuzmin, D., Rasskazova, S., & Timirgalin, A. (2018, September). Big data approach for geological study of the big region West Siberia. In *SPE Annual Technical Conference and Exhibition*. OnePetro.

Panetta, K. (2017). Gartner top 10 strategic technology trends for 2018. Smarter with gartner. Gartner, Inc.

Parshall, J. (2017). After years, 'big crew change' has passed, but learning, training challenges remain. *Journal of Petroleum Technology*, 69(7), 38–40.

Poddar, T. (2018, March). Digital twin bridging intelligence among man, machine and environment. In *Offshore Technology Conference Asia*. OnePetro.

Priyanka, E. B., Thangavel, S., Gao, X. Z., & Sivakumar, N. S. (2022). Digital twin for oil pipeline risk estimation using prognostic and machine learning techniques. *Journal of Industrial Information Integration*, 26, 100272.

Qi, Q., & Tao, F. (2018). Digital twin and big data towards smart manufacturing and Industry 4.0: 360 degree comparison. *IEEE Access*, 6, 3585–3593.

Roden, R. (2016, October). Seismic interpretation in the age of big data. In *2016 SEG International Exposition and Annual Meeting*. OnePetro.

Samimi, A. (2020). Risk management in oil and gas refineries. *Progress in Chemical and Biochemical Research*, 3(2), 140–146.

Semeraro, C., Lezoche, M., Panetto, H., & Dassisti, M. (2021). Digital twin paradigm: A systematic literature review. *Computers in Industry*, 130, 103469.

Settemsdal, S. O., & Bishop, B. (2019, September). When to go with cloud or edge computing in offshore oil and gas. In *SPE Offshore Europe Conference and Exhibition*. OnePetro.

Shirangi, M. G., Ettehadi, R., Aragall, R., Furlong, E., May, R., Dahl, T., & Thompson, C. (2020, February). Digital twins for drilling fluids: Advances and opportunities. In *IADC/SPE International Drilling Conference and Exhibition*. OnePetro.

Song, W., Li, F., Valero, M., & Zhao, L. (2019). Toward creating a subsurface camera. *Sensors*, 19(2), 301.

Souza, J. F. L., Santana, G. L., Batista, L. V., Oliveira, G. P., Roemers-Oliveira, E., & Santos, M. D. (2020). CNN prediction enhancement by post-processing for hydrocarbon detection in seismic images. *IEEE Access*, 8, 120447–120455.

Tao, F., Cheng, J., Qi, Q., Zhang, M., Zhang, H., & Sui, F. (2018). Digital twin-driven product design, manufacturing and service with big data. *The International Journal of Advanced Manufacturing Technology*, 94(9), 3563–3576.

Wanasinghe, T. R., Wroblewski, L., Petersen, B. K., Gosine, R. G., James, L. A., De Silva, O., ... Warrian, P. J. (2020). Digital twin for the oil and gas industry: Overview, research trends, opportunities, and challenges. *IEEE Access*, 8, 104175–104197.

Wu, Y., Zhang, K., & Zhang, Y. (2021). Digital twin networks: A survey. *IEEE Internet of Things Journal*, 8(18), 13789–13804.

Xia, F., Wang, W., Bekele, T. M., & Liu, H. (2017). Big scholarly data: A survey. *IEEE Transactions on Big Data*, 3(1), 18–35.

Zhang, H., Liu, Q., Chen, X., Zhang, D., & Leng, J. (2017). A digital twin-based approach for designing and multi-objective optimization of hollow glass production line. *IEEE Access*, 5, 26901–26911.

Zhang, H., Zhu, P., Liao, Z., & Li, Z. (2022). SaltISCG: Interactive salt segmentation method based on CNN and graph cut. *IEEE Transactions on Geoscience and Remote Sensing*, 60, 1–14.

4 Predictive Modelling Concepts in Petroleum Sector

Priyanka Singh and Noore Zahra

4.1 OVERVIEW

Recent advancements in the petroleum industries towards smart wells, real-time analytical visualization, and large-scale data interpretation are enhancing the optimization process as well as fulfilling the need for powerful, robust, and intelligent tools. The petroleum researchers and professionals are battling in their daily lives to solve highly complex and dynamic real-world issues, and high-stakes decision-making and planning tasks. For instance, several sensors are permanently placed in wellbores which capture very large amounts of data with vital importance. As such, relevant datasets need proper software to process data in a real-time or near-real-time manner to get the usefulness of exotic hardware tools in the petroleum sector. However, several processing and modelling software is available, but they barely scratch the surface of predictive modelling and analytical capabilities. Therefore, software with intelligent model systems is leveraged with only sustainable techniques to perform predictive analytics in real-time and decision-making capability. The intelligent systems must possess several features, hard (statistical) and soft (intelligent) computing methods to integrate artificial intelligence (AI) and machine learning (ML) techniques such as neural computing, fuzzy analysis, regression, and classification. Nevertheless, in recent years, these predictive modelling approaches have made various solid steps towards becoming more acceptable in the mainstream of the oil and gas industry to change its "black-box" image to a "transparent box". Nowadays, the exponential rise of AI in this industry covers high-level analysis and issues, from natural gas prediction for the next 15 years in the United States (US) (Al-Fattah and Startzman, 2001; Garcia and Mohaghegh, 2004), to managerial level decision-planning (Fletcher and Davis, 2002), as well as more mundane technical glitches that affect geoscientists/engineers such as well treatment (Mohaghegh et al., 2001a; Mohaghegh et al., 2001b), production engineering (Alimonti and Falcone, 2002;Weiss et al., 2002), reservoir characterization (Bhushan and Hopkinson, 2002; Finol et al., 2002; Mohaghegh et al., 2000; Mohaghegh et al., 1998), drilling (Balch et al., 2002), and many more. In this chapter, the statistical methods are discussed which are used for building artificial predictive models in the oil and gas industry.

DOI: 10.1201/9781003357872-4

4.2 STATISTICAL METHODS

The advancements in science and technology are the product of multiple investigations. For instance, Legendre and Gauss published an article related to the "least squares method" (Stigler, 1977), which is later introduced as linear regression, and it is effectively utilized to solve the issues in astronomy. Similarly, in 1936, Fisher introduced a linear discriminant analysis, then other researchers and statisticians promoted logistic regression as a substitute of regression for categorical data by the 1940s (Li and Wang, 2014). Linear regression is used to solve quantitative values, and logistic regression is used to solve qualitative values. Later, in the early 1970s, both linear and logistic regression cases were considered as a unique instance by Nelder and Wedderburn, which they further coined as generalized linear models. At the end of the 1980s the computing technology has improved significantly, and practical implementation was firstly demonstrated by Breiman, Friedman, Olshen, and Stone. This section reviews useful statistical techniques for multiple investigations by combining multiple data sources, identifying research requirements, choice of statistical methods on the statistical purpose and nature of data, and many other considerations. The statistical learning methods are classified as either supervised or unsupervised to understand the build of statistical models. Unsupervised learning methods are used to predict the output based on the inputs and unsupervised outputs, while supervised learning methods help predict the output based on one or more inputs and supervised outputs.

4.2.1 Parametric vs Non-Parametric Methods

In terms of determining whether they impose hypotheses on the distribution of data, methods to analyze numerical data are divided into two categories (Anderson, 1961). The parameters are the numbers for theoretical distributions; thus, parametric procedures, such as mean and standard deviation, are defined as those that apply distributional assumptions for data (Altman and Bland, 1999). The least squares regression, correlation tests, and *t* tests are some of the frequently used parametric techniques. These parametric methods are assumed to follow a normal distribution and studied observations are being uniformly spread (variance) either across the range or between the groups.

The alternative of parametric doesn't require data to follow a particular distribution pattern, rather they rely on ranks and order of observations. This alternative method is non-parametric that does not need to apply distributional theories of data and only follows rank-based procedures, instead of measurements. This non-parametric approach relates where analysis is to be done, not on property investigations (Altman and Bland, 1999). Therefore, rank methods almost have a similar ability as the *t* approach to discover an actual variation in case of larger samples, while also considering distributional data. However, transformation of skewed data could make it suitable for parametric analysis (Bland and Altman, 1996).

4.2.2 Regression and Classification

The response (or target) variables can be categorized into either qualitative or quantitative. The qualitative and quantitative variables are also known as categorical and continuous (numerical values) respectively. For example, the price of a house, stock prices, age, height, and a person's income belong to continuous variables. The categorical values are classified into different categories or classes (two or more); for example, a person's gender could be either male or female, whether the patient is diagnosed with cancer or not with different disease classes (Acute Lymphoblastic Leukemia and Acute Myelogenous Leukemia), or whether they default on a debt in a yes or no format. The problems with quantitative response variables are called regression problems, whereas the problems with quantitative responses are referred to as classification problems. The regression problems use the least squares linear regression technique and classification problems are tackled by logistic regression. However, a few methods such as boosting, and k-Nearest Neighbors (kNN) can be used in both cases of quantitative as well as qualitative response. The classification model is also known as a classifier that assigns an unclassified case to a predefined set of classes. The classifier is defined as a function $f(X)$ in which X denotes training samples and Y denotes a target variable that has to be predicted. These classifiers are used to solve real-world problems with more than two classes, such as disease classification (Austin et al., 2013; Badriyah et al., 2020; Gunarathne et al., 2017; Malik et al., 2019), face recognition (Sharif et al., 2016; Sharma et al., 2020; Zong and Huang, 2011), character recognition (Clanuwat et al., 2019; Coates et al., 2011; Liu and Fujisawa, 2008; Srivastava et al., 2019), text classification (Ikonomakis et al., 2005), Twitter user classification (Pennacchiotti and Popescu, 2011), and many more.

4.2.3 Performance Metrics for Classification and Regression

4.2.3.1 Performance Metrics for Classification—Confusion Matrix

A machine learning concept called a confusion matrix (Sammut and Webb, 2011) compiles data on both actual and predicted classifications made by a classification algorithm. The performance of a classifier is measured using a confusion matrix, which has two dimensions. One dimension is labelled "Actual", which indexes the class of an object, and the other is labelled "Predicted", which holds predicted values that the classifier predicts. Confusion matrices can be used to determine whether there is a large amount of class property overlap that might cause a model to become confused and make incorrect predictions for binary and multi-class classification issues. The fundamental confusion matrix for a

TABLE 4.1
Confusion Matrix Prototype

	Predicted	
Actual	TP	FP
	FN	TN

multi-class classification problem is shown in Table 4.1. The confusion matrix consists of the following four metrics:

i. True Positives (TP) refer to samples whose predicted and actual values are positive.
ii. True Negatives (TN) refer to predicted and actual values that are negative.
iii. False Positives (FP) denote those samples whose predicted value is positive but actual is negative. This is also known as a Type I error.
iv. False Negatives (FN) indicate those samples whose predicted value is negative but actual is positive. This is also known as a Type II error.

Using a confusion matrix, the number of performance metrics can be computed as follows:

Accuracy measures proportion of the correct predictions given by classification model. It is calculated as:

$$Accuracy = \frac{TP + TN}{TP + TN + FP + FN} \tag{4.1}$$

Precision is an accuracy measurement index that gives the ratio of true positive predictions and all positive prediction cases. It is calculated as:

$$Precision = \frac{TP}{TP + FP} \tag{4.2}$$

Recall measures ratio of true positive predictions with respect to actual positive samples and is also known as sensitivity. It is formulated as:

$$Recall = \frac{TP}{TP + FN} \tag{4.3}$$

Specificity is a reverse measure of recall which considers true negatives and all negative samples, and is defined in formula as:

$$Specificity = \frac{TN}{TN + FP} \tag{4.4}$$

F_1 *Score* is a traditional F score that represents the harmonic mean of recall and precision and is formulated as:

$$F = 2 \times \frac{Precision \times Recall}{Precision + Recall} \tag{4.5}$$

4.2.3.2 Performance Metrics for Regression Model

The reliability of regression models is evaluated by the following five parameters:

$$\text{Correlation coeeficient}, R = \frac{\Sigma\left(Y_{ms} - \bar{Y}_{ms}\right)\left(Y_{pd} - \bar{Y}_{pd}\right)}{\sqrt{\Sigma\left(Y_{ms} - \bar{Y}_{ms}\right)^2 . \Sigma\left(Y_{pd} - \bar{Y}_{pd}\right)^2}} \tag{4.6}$$

$$\text{Root Mean Squared Error, } RMSE = \sqrt{\frac{1}{n}\sum_{i=1}^{n}\left(Y_{pd} - Y_{ms}\right)^2} \tag{4.7}$$

$$\text{Mean Absolute Error, } MAE = \frac{1}{n}\sum_{i=1}^{n} | Y_{pd} - Y_{ms} | \tag{4.8}$$

$$\text{Mean Absolute Percentage Error, } MAPE = \frac{1}{n}\sum_{i=1}^{n}\left|\frac{Y_{ms} - Y_{pd}}{Y_{ms}}\right| \tag{4.9}$$

$$\text{Nash} - \text{Sutcliffe coefficient, } NSC = 1 - \left[\frac{\sum_{i=1}^{n}\left(Y_{ms} - Y_{pd}\right)^2}{\sum_{i=1}^{n}\left(Y_{ms} - \overline{Y}_{pd}\right)^2}\right] \tag{4.10}$$

Here, Y_{pd} refer to predicted values, $\overline{Y}_{pd}$ refer to the mean of predicted values, Y_{ms} is the actual value, $\overline{Y}_{ms}$ is the mean of actual values, and n is the total number of rows in data.

4.3 MACHINE LEARNING CONCEPTS

4.3.1 Background

The oil and gas industry is one of the biggest, complex, and most important sectors of the global economy. It involves a variety of activities like upstream, midstream, and downstream, which includes exploration, production, refining, transportation, and the consumption of oil. Oil is the primary fuel used worldwide, accounting for around one-third of the total energy consumed worldwide (International Energy Agency, 2009). Consequently, it ranks as one of the most important traded assets globally, petrochemicals are used in a wide variety of situations and daily essentials such as transportation, heating, and power (Inkpen and Moffett, 2011). The oil and gas firms must develop fresh ideas for enhancing supply operations because the demand is dynamic and always rising. If the price of fossil fuels keeps rising, fossil fuel enterprises need to develop new competence and reinforce procedures in order to increase efficiency and grow on their current capabilities. The traditional centralized strategy struggles to function because of its extremely dynamic characteristics, especially with the growth in the world market (Inkpen and Moffett, 2011). The industry must upgrade its technologies to discover suitable methods for handling and managing information, in order to improve operational effectiveness, cut operating costs, and increase profitability. For instance, to provide improved control and optimization of crude yield, real-time data streams continuously generated by sensors are being used (Kong and Ohadi, 2010).

The supply chain network of oil and gas is divided into three phases, each of which consists of a number of procedures. Some of these phases are upstream, midstream, and downstream. The exploration and production of crude are mostly involved in the upstream phase. The midstream is typically forgotten when detailing the supply chain phases, despite the fact that it involves storage and transit and covers both upstream and downstream. The downstream oversees refining oil and gas in addition to distribution of completed items. Every step of the supply chain is intricate, involving the use of operational, tactical, and strategic policy planning (Tsegha, 2013). To increase efficiency and personnel safety, robotics are utilized for damage control, drilling, and inspection in offshore areas (Shukla and Karki, 2016). Wireless networks with sensors enabled are applied to track and enhance productivity and recognize and resolve issues linked to health and safety (reza Akhondi et al., 2010). Radio-frequency identification (RFID) technology has benefitted safety, capital management, site supervision of oil rigs, security, and pipeline inspections (Felemban and Sheikh, 2013). It has become abundantly evident that digital technology has a significant impact on both business and society. With time, it has become clear that the "fourth industrial revolution" is the digital transformation, which is characterized by the merging of technologies like artificial intelligence, robots, and autonomous vehicles that shape the lines between biological, physical, and digital realms. Technologies based on artificial intelligence are attracting a lot of attention due to their quick response times and powerful generalization abilities (Evans, 2019). In a range of reservoir engineering challenges, machine learning has the potential to supplement and enhance traditional reservoir engineering methodologies (Anifowose et al., 2017). Much research uses sophisticated machine-learning methods for classification and regression problems, including Artificial Neural Networks (ANN), Fuzzy Logic (FL), Response Surface Model (RSM), and Supporting Vector Machines (SVM) (Ani et al., 2016).

In order to demonstrate the benefits of AI and ML techniques across different supply chain sectors in the oil and gas industry, this section reviewed the current research on the application of these techniques in the industry. This chapter examines how the petroleum industry has partnered with an intelligent numerical simulator to make work easier and boost productivity.

4.3.2 Machine Learning Methods

In the petroleum sector, numerous data formats are gathered by touching the surface and subsurface to know potentials of hydrocarbons. To fulfil the need of data collection, sensors are discovered to be vital for collecting data in a significant amount. These data need to be plotted and analyzed using technical analysis and interference. The result is predicted by ML techniques, listed in Figure 4.1, which also show relationships between the input variables. These sectors generate enormous amounts of data, and data correlation is a very challenging process, but the system's physical behaviour wouldn't be altered (Ali, 1994).

4.3.2.1 Linear Regression

The linear regression is one of the statistical techniques to detect linearity or measure closeness (correlation) between the variables. Forecasts for the world's oil production

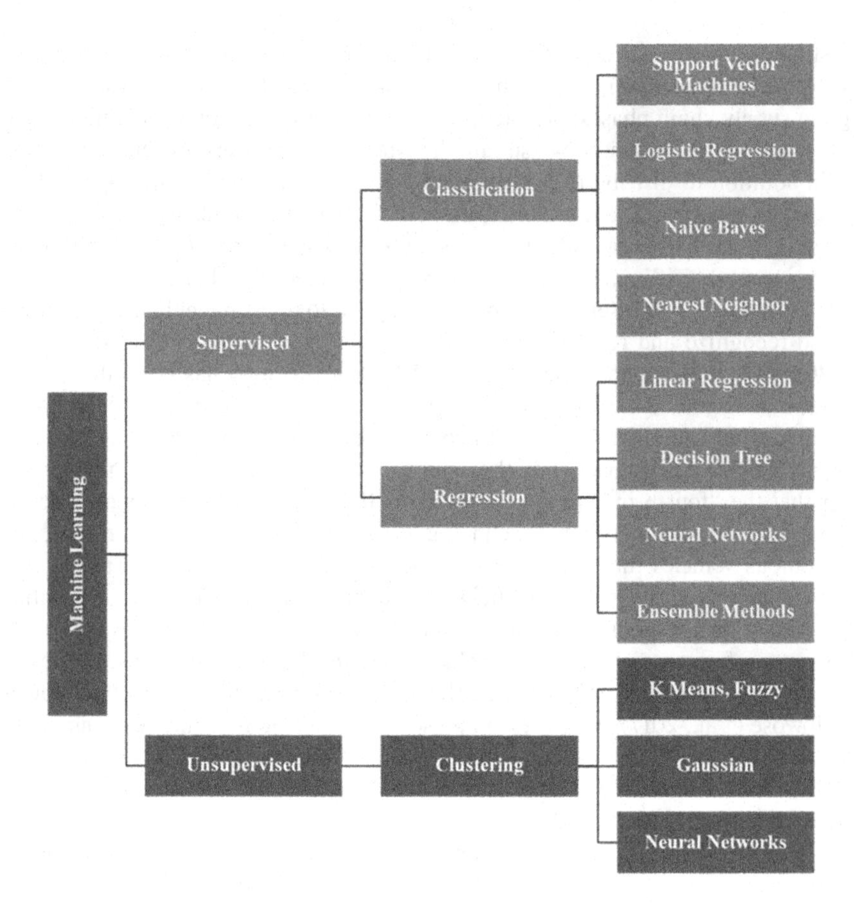

FIGURE 4.1 Hierarchy of machine learning algorithms.

are made using linear and non-linear regression models. As evidence, the linear regression involvement predicted that the world would produce 4,593 Mt of oil by 2020. The multiple linear regression (MLR) model is another form of linear regression which is mostly used for investigating the real well logging data.

MLR is a multivariate method to explain correlation among response (Y) and predictor (X_1, X_2, X_3, … ,X_n) variables. The MLR modelling is based on the equation as follows:

$$Y_t = X_t\beta + \varepsilon_t, \tag{4.11}$$

where X_t is a set of predictors at time, t, ε_t is an error by chance at time, t, Y_t refer to the response variable at time t, and β is the coefficient indicating the relationship between response and predictors. The model proved successful in recognizing patterns in the layers of oil and gas (Peng et al., 2016). Wang and Liu (2017) conducted a regression analysis on the elements that will affect crude oil's future economy.

4.3.2.2 Support Vector Machine

Support Vector Machine (SVM) is a ML method, first introduced by Vapnik (1999) to classify non-linear regression tasks. The various resources are available to illustrate its theoretical foundations (Arjmandzadeh et al., 2011; Vapnik, 1999). The working methodology of an SVM model is that a hyperplane(s) is constructed in a high dimensional feature space. The modified version of SVM for regression is known as Support Vector Regression (SVR), and the version proposed for a least squares method is Least Squares–SVM (LS–SVM) (Burges, 1998; Boswell, 2002). In the oil and gas sector, SVMs are widely used with reliable and significant outcomes (Hou and Wenfen, 2006; Taboada et al., 2007).

4.3.2.3 Strength Weakness Opportunities Threats (SWOT) Analysis

The two major phases of ML are experimentation and operationalization. The first one concentrates on efforts of data preprocessing, algorithm selection, validation of the model, and model evaluation. The latter one is the process of deploying models, followed by the consumption and monitoring of resilient, effective, and quantified services. Hajizadeh (2019) uses Strengths Weaknesses Opportunities Threats (SWOT) to communicate the gap between current ML developments and advances in the IT and E&P industries.

4.4 THE ARTIFICIAL NEURAL NETWORK AND ITS APPLICATION IN THE OIL AND GAS SECTOR

4.4.1 Introduction

Artificial Neural Network (ANN) is inspired by human cognitive functions and encircling all of them in machine behaviour. AI is completely based on algorithms, but it is difficult to categorize them in the domain of mathematical reasoning because of their adaption nature of insufficient resources and knowledge (Bravo et al., 2014). An Artificial Neural Network is an ML method which is represented in a graphical structure of nodes and their interconnected connections allow them to learn, train, and understand data. Therefore, ANN is the most powerful and traditional method (Wilamowski and Irwin, 2011) that is being applied to complex domains and real-world practical applications (McCulloch and Pitts, 1943; Prieto et al., 2016).

The simple and basic structure of ANN, illustrated in Figure 4.2, consists of vertical layers that include input layer as n neurons, hidden layers (one or more) as m neurons, and an output layer as o neurons. All the nodes will be interconnected with each other to regularize the flow of information in the network, and the activation function (linear or non-linear) is between the hidden and output layer. The activation functions are responsible for the abstract processing of neural networks by receiving the net input and transmitting an output. Therefore, ANN is proved to be efficient in complex computing because of its parallel distributed architecture owing to self-learning and generalization characteristics. The self-learning indicates that ANNs may adapt to various, and even changing, data settings without the need to "reshape" or "reprogramme", for algorithm development depending upon external conditions.

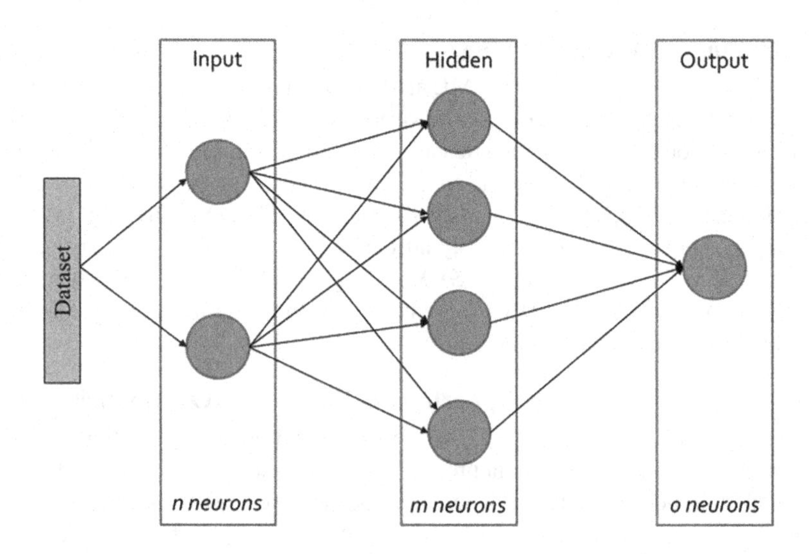

FIGURE 4.2 Example of ANN structure.

The generalization one is after is the self-learning process which is meant to illustrate how a new algorithm could be less sensitive to the quality of input because it can accommodate a wide range of potential inputs. This also offers a higher tolerance for data that is lacking in both quantity and quality. ANN is most frequently a ML technique used in the oil and gas industry to solve complex, non-linear problems that couldn't be solved by a linear one. The ANNs mostly belong to multi-layered and feedforward, where all nodes are connected vertically in a forward direction and data at input layer flows to the next layers setting at traversed levels (Ashena and Thonhauser, 2015). The neural network application areas in the petroleum sector are drill bit diagnosis, well performance prediction, seismic pattern recognition, sandstone lithofacies identification, and increased gas well production (Ali, 1994). The ANN model aids in predicting pipeline conditions and provides operators the ability to evaluate and forecast pipeline conditions. Tabesh et al. (2009) analyze the mechanical reliability and predicted pipe failure rate using ANN and other approaches. The back propagation algorithm is another useful algorithm, introduced in 1986 by Rumelhart et al. (1986), that agrees adjustment of threshold and network weights constantly to-and-fro to reduce the sum of a squared error.

4.4.2 Generic Development of ANN Models

This section provides a brief discussion over important phases that an ANN researcher and/or developer should consider while developing ANN models.

i. The proper approach is required to ensure that input variables are relevant and independent so that the ANN's accuracy might not be affected (Gholizadeh et al., 2011; Kasun et al., 2016). For this, dimensionality reduction in inputs is a necessary step which could be carried out by feature selection and feature extraction. Some of them are sparse random projections, autoencoder, and linear correlation analysis (Kasun et al., 2016).

ii. After input definition, the next mandatory step is to define the shared amount of input into training, validation, and testing sets. The traditional input sharing amounts are 50–25–25, 60–20–20, 70–15–15, or 80–10–10 (Abambres, 2018).

iii. If concentration of training samples are heterogeneous in nature, then the training efficiency could be diminished. Therefore, input normalization techniques are required such as Pu and Lachtermacher and Fuller (1995); Pu and Mesbahi (2006); and Tohidi and Sharifi (2015).

iv. The transfer functions must be deployed between hidden and output layers. The commonly used transfer functions are sigmoid (logistic), hyperbolic tangent, bilinear, sinusoid and radial basis, positive saturating linear, identity–logistic, identity, and bipolar functions. Afterwards, output normalization could be done to ignore numerical glitches during training (McCulloch and Pitts, 1943; Prieto et al., 2016).

v. Network connectivity arrangement should be defined based on the behaviour of hidden layer's nodes, even in multilayer feedforward ANN structure.

4.4.3 Applications in the Oil and Gas Industry

This section discusses applications of state-of-art technologies in different phases of the petroleum sector.

4.4.3.1 Exploration

Hydrocarbon exploration is a hazardous task because personnel need to touch the subsurface for drilling and hydrocarbon exploitation. Earlier, in the 21st century, a very little amount of 2D seismic data was used to locate drilling sites based on their subsurface mapping with a likelihood success ratio of 1:7. More data was gathered over time in each of the lease curves, and AI and ML methods are used to analyze these voluminous datasets in India and the world. The drillable prospects are converted using ML methods and this improves the success ratio to 1:3. However, ANN models are frequently used to improve the target's potential, size, and hydrocarbon's concentration. The implementation idea of AI is responsible for infiltrating different phases of this sector, for instance intelligent processing drilling, development, and so on. However, various researchers and research groups developed realistic AI-based technologies in research and production. In terms of lowering exploration risks and increasing success rates of exploration wells, the use of the ANN technique in the field of exploration has already yielded positive results (Pandey et al., 2017). To understand the characteristics of data, unsupervised learning could be useful. A framework proposed by Kumar (2019) was successful for the Shales project, since it could manage large-scale datasets. Another innovation is a recurrent neural network that created artificial well log data accurately and at a reasonable price from existing well log data (Zhang et al., 2018). Diersen et al.

(2011) used AI to reduce the labour-intensive processes involved in processing and analyzing seismic full wave tomography.

4.4.3.2 Reservoir

In recent years, ANN has been used to estimate reservoir parameters like porosity and permeability. The study of forecasting reservoir fluid properties can be carried out by applying several machine learning techniques such Adaptive Boosting, K-Nearest Neighbours (KNN), Kernel Ridge Regression (KRR), Support Vector Regression (SVR), and Collaborative Filtering. In the reservoir study by Onwuchekwa (2018), it was discovered that collaborative filtering was made for recommendation systems of a consumer's product, and this worked effectively for such. Teixeira and Secchi (Teixeira and Secchi, 2019) used optimization techniques for identifying the ideal control to boost overall oil production. The parametric analysis could be done by assessing several machine learning methods to forecast permeability, wireline data, and seismic properties. The SVM performed better than other methods at predicting permeability (Anifowose et al., 2017). Extreme Gradient Boosting was used by Anifowose et al. (2019) to build an intelligent model that forecasts reservoir response through injector wells. A 20-model ensemble water flood, a homogeneous reservoir water flood, a channelized reservoir water flood, and a CO_2 flood in a heterogeneous reservoir with challenging terrain were the five examples selected by Nwachukwu et al. (2018).

4.4.3.3 Drilling

Drilling phase is having several issues, including borehole instability, high torque, bit wear, stick sleep vibrations, etc., which can be addressed through ML techniques (Noshi and Schubert, 2018). Aliouane and Ouadfeul (2014) proposed the preparation of poisson's ratio map using ML for determining the direction of drilling and the features of the rocks. Castiñeira et al. (2018) used the ML approach to assess the drilling data's quality, gather critical data, and forecast non-productive time. This approach was helpful in reducing labour costs, invested for quality-checking of massive amounts of drilling data. According to Bhandari et al. (2015), the Bayesian network can be used to analyze risks and predict failures in the offshore industries. An ML technique can be used to acquire information such as estimates of abrasiveness, alternative bit or rig equipment upgrades, and anticipated bit wear (Dunlop et al., 2011).

4.5 CONCEPTS OF DEEP NEURAL NETWORKS

4.5.1 Introduction

In recent decades, deep neural networks (DNNs) are serving as the base for many modern applications of AI (LeCun et al., 2015) ranging from complex to extreme complex problems such as image recognition (Krizhevsky et al., 2017), speech recognition (Deng et al., 2013), cancer detection (Esteva et al., 2017), self-driving cars (Chen et al., 2015) and many more. DNNs' tremendous performance is because of their ability to extract complex and advanced features from raw data by utilizing statistical learning over a vast quantity of data to derive an effective pattern, like humans can. The demonstration of its working phenomena, for instance, on image data processing is explained here. Firstly, the pixels of an image are given as an input to the input layer of neural networks, whose output is indicated as the presence of various low-level features in the input image, such as lines and edges at the next layers. These low-level features are then merged with available higher-level features to form a set of shapes; for example, lines are formed into shapes. A final step of the network is to provide a probability that these high-level features consist of certain objects or visuals. This deep hierarchical characteristic allows DNNs to execute various complex tasks with good performance ability. While many DNN processes have relied heavily on general-purpose computing engines, particularly graphics processing units (GPUs), there is growing interest in offering more specialized acceleration of DNN computation.

There are two main types of networks to handle the input: feedforward and recurrent. In feedforward networks, all computation is carried out on a previous layer's outputs in a series of actions and then, the final processing produces output. As such, DNNs do not have any memory in their network and also, their output is the same as an input, despite the consequences of the order in which the network has previously received inputs. While this is the case, long short-term memory networks (LSTMs) (Hochreiter and Schmidhuber, 1997), a form of recurrent neural networks (RNNs), with internal memory enabled long-term dependencies to impact the output. Within these networks, outputs of intermediate processing layers are stored internally and later, they are used as inputs to additional operations in connection with processing of a subsequent input. However, to make the computing more efficient, DNN later adopted the computation structuring as a convolution and these multiple convolution-based layers form a new variant of DNN, which is known as a convolutional neural network (CNN).

4.5.2 DNN Models

In this section, popular DNN models with different network architecture will be briefed to understand the trends and variations, which is required prior to implementing the DNN engine. One of the first CNN models was introduced in 1989 as LeNet (LeCun et al., 1989) whose job is to classify digits in grayscale images of 28 × 28 pixel resolution. Its popular version is LeNet-5 that consists of two convolutions and two fully connected layers (LeCun et al., 1998), and each convolution uses six filters per layer with a 5 × 5 filter size. After each convolution, a 2 × 2 average pooling is used, and sigmoid function for nonlinearity is applied. However, LeNet needs 341,000 multiply-and-accumulates (MACs) and 60,000 weights overall for each image. LeNet's first successful application is ATMs where it is deployed to identify and classify digits for check deposits.

Another popular and efficient DNN model is AlexNet (Krizhevsky et al., 2017), which comprises three fully connected and five convolution layers. At each convolution, 96 to 384 filters (size ranges from 3 × 3 to 11 × 11) with 3 to 256 channels.

The red, green, and blue elements of the input image are represented by the three channels of the filter in the first layer. In each layer, a ReLU nonlinearity is utilized. The outputs of layers 1, 2, and 5 are subjected to a max pooling of 3 × 3. At the top layer of the network, a stride of four is used to minimize computation. The fact that there are more weights and various shapes between layers in AlexNet than in LeNet is one of the key differences between the two networks. The 96 output channels of the first layer are divided into two groups of 48 input channels for the second layer, resulting in a second convolution layer with only 48 channels to diminish the computation and weights. This approach is applicable on the weights in the next fourth and fifth layers. Therefore, a 227 × 227 input image needs 724 million MACs and 61 million weights in total by AlexNet to process that single image.

The VGG-16 (Simonyan and Zisserman, 2014) is another DNN model which is deeper by including 13 layers of convolution and 3 fully connected layers, in total 16 layers. This model needs 138 million weights and 15.5G MACs to classify 224 × 224 resolution of an image. Apart from VGG-16, VGG has one more variant; i.e., VGG-19 that provides a top-5 error rate, which is 0.1% lower than VGG-16's with respect to the cost of 1.27 times more MACs.

Residual Net (ResNet) (He et al., 2016) uses "residual connections" to make a deeper network of 34 layers or more. This was the first DNN in ImageNet challenge to exceed a human level accuracy that had a top-5 error rate under 5%. The "shortcut" module introduced by residual net has an identity link that enables skipping the weight of convolution layers. ResNet-50 consists of one convolution and 16 "shortcut" layers and needs 3.9G MACs and 25.5 million weights to process each image.

4.5.3 Model Evaluation Metrics

The following metrics should be taken into account while assessing a specific DNN model's characteristics:

i. The model's precision as measured by the top-5 errors on datasets like ImageNet. Additionally, it should be stated what kind of augmentation was utilized (multiple crops, ensemble models, etc.).
ii. The network architecture of the model, including the number of layers, filter sizes, and channels, should be disclosed.
iii. The amount of weights has an impact on the model's storage needs and needs to be reported. In order to reflect the theoretical minimal storage requirements, it is best to record the number of nonzero weights.
iv. It is important to indicate the number of MACs that must be completed because it can provide some insight into the quantity of activities and possibilities.

4.6 CASE STUDY

In this case study by Sousa et al. (2019), the supply network, depicted in Figure 4.3, with a bullwhip effect in oil and gas is being investigated here. The suppliers need to

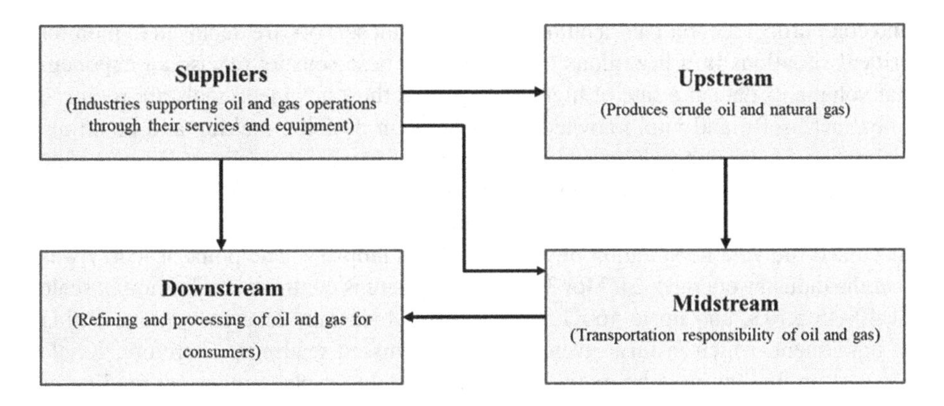

FIGURE 4.3 Levels of oil and gas supply chain.

adjust their workload to meet with the peaks of large demand (Kadivar and Shirazi, 2018) and cope with the cyclic variations of high and low demand (de Oliveira Pacheco et al., 2017). As such, it challenges industrial management, and raises sustainability, maintenance, and technical issues (Sousa et al., 2019), promoted through excess and underproduction. The concept of the bullwhip effect was introduced in the fourth industrial revolution (Müller and Däschle, 2018; Stock and Seliger, 2016), similar to multi-level supply networks. Therefore, usage of artificial neural networks (ANN) may provide better tools for the community. The questions formulated by Sousa were to address the occurrence of the bullwhip effect, whether suppliers exhibit highest demand variability in network, whether small-scale companies are more susceptible with highest demand variability than large-scale companies, and whether ANN will be useful in forecasting the bullwhip effect. Therefore, as per the scope of the paper, the use of ANN in the bullwhip effect is more concentrated than others. The methodology planned for this investigation includes a detailed and deeper literature review, data collection planning through survey, and ANN development. The literature review is conducted to identify the behavioural and systemic factors having an impact on the bullwhip effect of the supply chain network. The next phase is to collect and analyze data related to production and financial information from the New York Stock Exchange (Azhar, 2013), reports of national authorities, projects, stock markets, and businesses. Among these companies, those companies were selected for investigation which were having direct involvement and then, verified whether their dataset is publicly available in a frequent period. The companies were arranged as per their levels in the supply chain and as demonstrated and described in Figure 4.1. After the data gathering and analysis phase, the development of an ANN-based demand forecasting model is the next stage of the research programme. The selection of ANN depends upon the nature of the problem, resources, and the researchers' proficiency. Therefore, different neural networks will be used, and a thorough comparative analysis will be done to minimize prediction error.

Another case study discussed here is about the research carried out by Kejela at el. (2014) on real-world drilling sensor data. With the recent advancements of sensor and computing technologies, a number of significant sensors are deployed to monitor critical situations in a hazardous environment. These sensors release an exponential volume of data at a rate of high velocity, and thus, advanced tools are required to extract useful and vital knowledge that helps in decision-making and planning, estimation of missing values, and replacement of incorrect readings. By using the example of surface equipment instrumentation, subjected to - 60°C to 180°C temperature and down-hole sensors temperature is up to 260°C, Florence et al. (2012) discussed the volatile situation of the oil and gas industry. The point of worry was that the industry operates 24/7 for 365 days and there is continuous vibration in scale of 40–45 RMS, and up to 160G shock loads, which may lead to malfunctioning of equipment, which in turn gives incorrect or missed readings. Therefore, Kejela focused on developing a predictive analytics model to replace incorrect reading or forecast missing values accurately. For this, the complete sensor data was partitioned into training and test samples of 504,389 and 10,351 observations, respectively. This dataset consists of 135 columns, out of which 1 was treated as response and the remaining were predictors. Then, the author found the top 10 predictors using Gradient Boosted Model (GBM), which were relevant in the prediction of gear oil temperature. Then, GBM and the Generalized Linear Model (GLM) algorithm was performed on training sets. Firstly, the regression problem was faced by forecasting the gear oil temperature, and then a classification problem was solved to predict whether the Inline Blow Out Preventer (IBOP) was opened or closed. The performance of these models was evaluated by performance metrics such as a confusion matrix. As a result, both GLM and GBM models have given similar accuracy, and this concluded that the predictive model is safe, stable, and can consistently predict missing or incorrect readings.

BIBLIOGRAPHY

Abambres, M., Rajana, K., Tsavdaridis, K. D., & Ribeiro, T. P. (2018). Neural network-based formula for the buckling load prediction of I-section cellular steel beams. *Computers*, *8*(1), 2.

Al-Fattah, S. M., & Startzman, R. A. (2001). Predicting natural gas production using artificial neural network. In *SPE Hydrocarbon Economics and Evaluation Symposium*. OnePetro.

Ali, J. K. (1994). Neural networks: A new tool for the petroleum industry?. In *European Petroleum Computer Conference*. OnePetro.

Alimonti, C., & Falcone, G. (2002, September). Knowledge discovery in databases and multiphase flow metering: The integration of statistics, data mining, neural networks, fuzzy logic, and ad hoc flow measurements towards well monitoring and diagnosis. In *SPE Annual Technical Conference and Exhibition*. OnePetro.

Aliouane, L., & Ouadfeul, S. A. (2014). Sweet spots discrimination in shale gas reservoirs using seismic and well-logs data. A case study from the Worth Basin in the Barnett Shale. *Energy Procedia*, *59*, 22–27.

Altman, D. G., & Bland, J. M. (1999). Statistics notes variables and parameters. *BMJ*, *318*(7199), 1667.

Anderson, N. H. (1961). Scales and statistics: Parametric and nonparametric. *Psychological Bulletin*, *58*(4), 305.

Ani, M., Oluyemi, G., Petrovski, A., & Rezaei-Gomari, S. (2016). Reservoir uncertainty analysis: The trends from probability to algorithms and machine learning. In *SPE Intelligent Energy International Conference and Exhibition*. OnePetro.

Anifowose, F., Abdulraheem, A., & Al-Shuhail, A. (2019). A parametric study of machine learning techniques in petroleum reservoir permeability prediction by integrating seismic attributes and wireline data. *Journal of Petroleum Science and Engineering, 176*, 762–774.

Anifowose, F. A., Labadin, J., & Abdulraheem, A. (2017). Ensemble machine learning: An untapped modeling paradigm for petroleum reservoir characterization. *Journal of Petroleum Science and Engineering, 151*, 480–487.

Arjmandzadeh, A., Effati, S., & Zamirian, M. (2011). Interval support vector machine in regression analysis. *The Journal of Mathematics and Computer Science, 2*, 565–571.

Ashena, R., & Thonhauser, G. (2015). Application of artificial neural networks in geoscience and petroleum industry. In *Artificial Intelligent Approaches in Petroleum Geosciences* (pp. 127–166). Springer, Cham.

Austin, P. C., Tu, J. V., Ho, J. E., Levy, D., & Lee, D. S. (2013). Using methods from the data-mining and machine-learning literature for disease classification and prediction: A case study examining classification of heart failure subtypes. *Journal of Clinical Epidemiology, 66*(4), 398–407.

Aydin, G. Ö. K. H. A. N. (2014). Production modeling in the oil and natural gas industry: An application of trend analysis. *Petroleum Science and Technology, 32*(5), 555–564.

Azhar, M. (2013). *The Study of the Bullwhip Effect in the Oil and Gas Industry*. Oklahoma State University.

Badriyah, T., Sakinah, N., Syarif, I., & Syarif, D. R. (2020). Machine learning algorithm for stroke disease classification. In *2020 International Conference on Electrical, Communication, and Computer Engineering (ICECCE)* (pp. 1–5). IEEE.

Balch, R. S., Hart, D. M., Weiss, W. W., & Broadhead, R. F. (2002, April). Regional data analysis to better predict drilling success: Brushy Canyon formation, Delaware Basin, New Mexico. In *SPE/DOE Improved Oil Recovery Symposium*. OnePetro.

Bhandari, J., Abbassi, R., Garaniya, V., & Khan, F. (2015). Risk analysis of deepwater drilling operations using Bayesian network. *Journal of Loss Prevention in the Process Industries, 38*, 11–23.

Bhushan, V., & Hopkinson, S. C. (2002, October). A novel approach to identify reservoir analogues. In *European Petroleum Conference*. OnePetro.

Bland, J. M., & Altman, D. G. (1996). Statistics notes: Transforming data. *BMJ, 312*(7033), 770.

Boswell, D. (2002). *Introduction to Support Vector Machines*. Department of Computer Science and Engineering University of California San Diego, Caltech.

Bravo, C., Saputelli, L., Rivas, F., Pérez, A. G., Nikolaou, M., Zangl, G., ... Nunez, G. (2014). State of the art of artificial intelligence and predictive analytics in the E&P industry: A technology survey. *Spe Journal, 19*(4), 547–563.

Burges, C. J. (1998). A tutorial on support vector machines for pattern recognition. *Data Mining and Knowledge Discovery, 2*(2), 121–167.

Castiñeira, D., Toronyi, R., & Saleri, N. (2018, April). Machine learning and natural language processing for automated analysis of drilling and completion data. In *SPE Kingdom of Saudi Arabia Annual Technical Symposium and Exhibition*. OnePetro.

Chen, C., Seff, A., Kornhauser, A., & Xiao, J. (2015). Deepdriving: Learning affordance for direct perception in autonomous driving. In *Proceedings of the IEEE International Conference on Computer Vision* (pp. 2722–2730).

Coates, A., Carpenter, B., Case, C., Satheesh, S., Suresh, B., Wang, T., ... Ng, A. Y. (2011, September). Text detection and character recognition in scene images with unsupervised feature learning. In *2011 International Conference on Document Analysis and Recognition* (pp. 440–445). IEEE.

Clanuwat, T., Lamb, A., & Kitamoto, A. (2019). Kuronet: Pre-modern Japanese kuzushiji character recognition with deep learning. In *2019 International Conference on Document Analysis and Recognition (ICDAR)* (pp. 607–614). IEEE.

de Oliveira Pacheco, E., Cannella, S., Lüders, R., & Barbosa-Povoa, A. P. (2017). Order-up-to-level policy update procedure for a supply chain subject to market demand uncertainty. *Computers & Industrial Engineering, 113*, 347–355.

Deng, L., Li, J., Huang, J. T., Yao, K., Yu, D., Seide, F., ... Acero, A. (2013, May). Recent advances in deep learning for speech research at Microsoft. In *2013 IEEE International Conference on Acoustics, Speech and Signal Processing* (pp. 8604–8608). IEEE.

Diersen, S., Lee, E. J., Spears, D., Chen, P., & Wang, L. (2011). Classification of seismic windows using artificial neural networks. *Procedia Computer Science, 4*, 1572–1581.

Dunlop, J., Isangulov, R., Aldred, W. D., Sanchez, H. A., Flores, J. L., Herdoiza, J. A., ... Luppens, J. C. (2011, March). Increased rate of penetration through automation. In *SPE/IADC Drilling Conference and Exhibition*. OnePetro.

Esteva, A., Kuprel, B., Novoa, R. A., Ko, J., Swetter, S. M., Blau, H. M., & Thrun, S. (2017). Dermatologist-level classification of skin cancer with deep neural networks. *Nature, 542*(7639), 115–118.

Evans, S. J. (2019). How digital engineering and cross-industry knowledge transfer is reducing project execution risks in oil and gas. In *Offshore Technology Conference*. OnePetro.

Felemban, E., & Sheikh, A. A. (2013). RFID for oil and gas industry: Applications and challenges. *International Journal of Engineering and Innovative Technology, 3*, 80–85.

Finol, J., Romero, C., & Romero, P. (2002, September). An intelligent identification method of fuzzy models and its applications to inversion of NMR logging data. In *SPE Annual Technical Conference and Exhibition*. OnePetro.

Fletcher, A., & Davis, J. P. (2002, October). Decision-making with incomplete evidence. In *SPE Asia Pacific Oil and Gas Conference and Exhibition*. OnePetro.

Florence, F. R., & Burks, J. (2012, May). New surface and down-hole sensors needed for oil and gas drilling. In *2012 IEEE International Instrumentation and Measurement Technology Conference Proceedings* (pp. 670–675). IEEE.

Garcia, A., & Mohaghegh, S. D. (2004, September). Forecasting US natural gas production into year 2020: A comparative study. In *SPE Eastern Regional Meeting*. OnePetro.

Gholizadeh, S., Pirmoz, A., & Attarnejad, R. (2011). Assessment of load carrying capacity of castellated steel beams by neural networks. *Journal of Constructional Steel Research, 67*(5), 770–779.

Gunarathne, W. H. S. D., Perera, K. D. M., & Kahandawaarachchi, K. A. D. C. P. (2017). Performance evaluation on machine learning classification techniques for disease classification and forecasting through data analytics for chronic kidney disease (CKD). In *2017 IEEE 17th International Conference on Bioinformatics and Bioengineering (BIBE)* (pp. 291–296). IEEE.

Hajizadeh, Y. (2019). Machine learning in oil and gas; a SWOT analysis approach. *Journal of Petroleum Science and Engineering, 176*, 661–663.

He, K., Zhang, X., Ren, S., & Sun, J. (2016). Deep residual learning for image recognition. In *Proceedings of the IEEE Conference on Computer Vision and Pattern Recognition June 27* 2016 *to June 30* 2016, *Las Vegas, NV, USA* (pp. 770–778).

Hou, J., & Wenfen, H. (2006). Novel approach to predict potentiality of enhanced oil recovery. In *Intelligent Energy Conference and Exhibition*. OnePetro.

Hochreiter, S., & Schmidhuber, J. (1997). Long short-term memory. *Neural Computation, 9*(8), 1735–1780.

Ikonomakis, M., Kotsiantis, S., & Tampakas, V. (2005). Text classification using machine learning techniques. *WSEAS Transactions on Computers, 4*(8), 966–974.

Inkpen, A. C., & Moffett, M. H. (2011). *The Global Oil & Gas Industry: Management, Strategy & Finance*. PennWell Books, Tulsa, Oklahoma.

International Energy Agency. (2009). *World Energy Outlook* (p. 17). Paris: OECD/IEA.

Kadivar, M., & Shirazi, M. A. (2018). Analyzing the behavior of the bullwhip effect considering different distribution systems. *Applied Mathematical Modelling, 59*, 319–340.

Kasun, L. L. C., Yang, Y., Huang, G. B., & Zhang, Z. (2016). Dimension reduction with extreme learning machine. *IEEE Transactions on Image Processing, 25*(8), 3906–3918.

Kejela, G., Esteves, R. M., & Rong, C. (2014, December). Predictive analytics of sensor data using distributed machine learning techniques. In *2014 IEEE 6th International Conference on Cloud Computing Technology and Science* (pp. 626–631). IEEE.

Kong, X., & Ohadi, M. M. (2010). Applications of micro and nano technologies in the oil and gas industry-an overview of the recent progress. In *Abu Dhabi International Petroleum Exhibition and Conference*. OnePetro.

Krizhevsky, A., Sutskever, I., & Hinton, G. E. (2017). Imagenet classification with deep convolutional neural networks. *Communications of the ACM, 60*(6), 84–90.

Kumar, A. (2019, April). A machine learning application for field planning. In *Offshore Technology Conference*. OnePetro.

Lachtermacher, G., & Fuller, J. D. (1995). Back propagation in time-series forecasting. *Journal of Forecasting, 14*(4), 381–393.

LeCun, Y., Bengio, Y., & Hinton, G. (2015). Deep learning. *Nature, 521*(7553), 436–444.

LeCun, Y., Bottou, L., Bengio, Y., & Haffner, P. (1998). Gradient-based learning applied to document recognition. *Proceedings of the IEEE, 86*(11), 2278–2324.

Liu, C. L., & Fujisawa, H. (2008). Classification and learning methods for character recognition: Advances and remaining problems. In *Machine Learning in Document Analysis and Recognition*(Eds Simone Marinai & Hiromichi Fujisawa) (pp. 139–161). Springer, Berlin, Heidelberg.

Li, C., & Wang, B. (2014). *Fisher Linear Discriminant Analysis*. CCIS Northeastern University, Boston, Massachusetts.

Malik, S., Kanwal, N., Asghar, M. N., Sadiq, M. A. A., Karamat, I., & Fleury, M. (2019). Data driven approach for eye disease classification with machine learning. *Applied Sciences, 9*(14), 2789.

McCulloch, W. S., & Pitts, W. H. (1943). A logical calculus of the ideas immanent in nervous activity. *Bulletin of Mathematical Biology, 1*, 1–1.

Mohaghegh, S. D., Goddard, C., Popa, A., Ameri, S., & Bhuiyan, M. (2000, October). Reservoir characterization through synthetic logs. In *SPE Eastern Regional Meeting*. OnePetro.

Mohaghegh, S., Mohamad, K., Popa, A., Ameri, S., & Wood, D. (2001a). Performance drivers in restimulation of gas-storage wells. *SPE Reservoir Evaluation & Engineering, 4*(6), 536–542.

Mohaghegh, S., Platon, V., & Ameri, S. (2001b). Intelligent systems application in candidate selection and treatment of gas storage wells. *Journal of Petroleum Science and Engineering, 31*(2–4), 125–133.

Mohaghegh, S., Richardson, M., & Ameri, S. (1998, November). Virtual magnetic imaging logs: Generation of synthetic MRI logs from conventional well logs. In *SPE Eastern Regional Meeting*. OnePetro.

Müller, J. M., & Däschle, S. (2018). Business model innovation of industry 4.0 solution providers towards customer process innovation. *Processes, 6*(12), 260.

Noshi, C. I., & Schubert, J. J. (2018, October). The role of machine learning in drilling operations; a review. In *SPE/AAPG Eastern Regional Meeting*. OnePetro.

Nwachukwu, A., Jeong, H., Pyrcz, M., & Lake, L. W. (2018). Fast evaluation of well placements in heterogeneous reservoir models using machine learning. *Journal of Petroleum Science and Engineering, 163*, 463–475.

Onwuchekwa, C. (2018, August). Application of machine learning ideas to reservoir fluid properties estimation. In *SPE Nigeria Annual International Conference and Exhibition*. OnePetro.

Pandey, R. K., Kakati, H., & Mandal, A. (2017). Thermodynamic modeling of equilibrium conditions of CH4/CO2/N2 clathrate hydrate in presence of aqueous solution of sodium chloride inhibitor. *Petroleum Science and Technology*, *35*(10), 947–954.

Peng, Z., Yang, H., Pan, H., & Ji, Y. (2016, August). Identification of low resistivity oil and gas reservoirs with multiple linear regression model. In *2016 12th International Conference on Natural Computation, Fuzzy Systems and Knowledge Discovery (ICNC-FSKD)* (pp. 529–533). IEEE.

Pennacchiotti, M., & Popescu, A. M. (2011). A machine learning approach to twitter user classification. In *Proceedings of the International AAAI Conference on Web and Social Media,* Barcelona (Spain), (Vol. 5, No. 1, pp. 281–288).

Prieto, A., Prieto, B., Ortigosa, E. M., Ros, E., Pelayo, F., Ortega, J., & Rojas, I. (2016). Neural networks: An overview of early research, current frameworks and new challenges. *Neurocomputing*, *214*, 242–268.

Pu, Y., & Mesbahi, E. (2006). Application of artificial neural networks to evaluation of ultimate strength of steel panels. *Engineering Structures*, *28*(8), 1190–1196.

reza Akhondi, M., Talevski, A., Carlsen, S., & Petersen, S. (2010, April). Applications of wireless sensor networks in the oil, gas and resources industries. In *2010 24th IEEE International Conference on Advanced Information Networking and Applications* (pp. 941–948). IEEE.

Rumelhart, D. E., Hinton, G. E., & Williams, R. J. (1986). Learning representations by back-propagating errors. *Nature*, *323*(6088), 533–536.

Sammut, C., & Webb, G. I. (Eds.). (2011). *Encyclopedia of Machine Learning*. Springer Science & Business Media. Springer New York. DOI 10.1007/978-0-30164-0.

Sharif, M., Bhagavatula, S., Bauer, L., & Reiter, M. K. (2016). Accessorize to a crime: Real and stealthy attacks on state-of-the-art face recognition. In *Proceedings of the 2016 Acm Sigsac Conference on Computer and Communications Security* October 24–28, 2016, Vienna, Austria (pp. 1528–1540).

Sharma, S., Bhatt, M., & Sharma, P. (2020, June). Face recognition system using machine learning algorithm. In *2020 5th International Conference on Communication and Electronics Systems (ICCES)* (pp. 1162–1168). IEEE.

Shukla, A., & Karki, H. (2016). Application of robotics in offshore oil and gas industry—A review part II. *Robotics and Autonomous Systems*, *75*, 508–524.

Sousa, A. L., Matos, H. A., & Guerreiro, L. P. (2019). Preventing and removing wax deposition inside vertical wells: A review. *Journal of Petroleum Exploration and Production Technology*, *9*(3), 2091–2107.

Sousa, A. L., Ribeiro, T. P., Relvas, S., & Barbosa-Póvoa, A. (2019). Using machine learning for enhancing the understanding of bullwhip effect in the oil and gas industry. *Machine Learning and Knowledge Extraction*, *1*(3), 57.

Srivastava, S., Priyadarshini, J., Gopal, S., Gupta, S., & Dayal, H. S. (2019). Optical character recognition on bank cheques using 2D convolution neural network. In *Applications of Artificial Intelligence Techniques in Engineering* (pp. 589–596). Springer, Singapore.

Stigler, S. M. (1977). An attack on Gauss, published by Legendre in 1820. *Historia Mathematica*, *4*(1), 31–35.

Stock, T., & Seliger, G. (2016). Opportunities of sustainable manufacturing in industry 4.0. *Procedia CIRP*, *40*, 536–541.

Tabesh, M., Soltani, J., Farmani, R., & Savic, D. (2009). Assessing pipe failure rate and mechanical reliability of water distribution networks using data-driven modeling. *Journal of Hydroinformatics*, *11*(1), 1–17.

Taboada, J., Matías, J. M., Ordóñez, C., & García, P. J. (2007). Creating a quality map of a slate deposit using support vector machines. *Journal of Computational and Applied Mathematics*, *204*(1), 84–94.

Teixeira, A. F., & Secchi, A. R. (2019). Machine learning models to support reservoir production optimization. *IFAC-PapersOnLine*, *52*(1), 498–501.

Tohidi, S., & Sharifi, Y. (2015). Inelastic lateral-torsional buckling capacity of corroded web opening steel beams using artificial neural networks. *The IES Journal Part A: Civil & Structural Engineering*, *8*(1), 24–40.

Tsegha, E. (2013). Assessing the challenges and opportunities in the oil and gas industry. *Academic Journal of Interdisciplinary Studies*, *2*(12), 129.

Vapnik, V. (1999). *The Nature of Statistical Learning Theory.* Springer Science & Business Media, Springer New York, NY, ISBN978-0-387-98780-4

Wang, K., & Liu, H. (2017). Regression analysis of influencing factors on the future price of crude oil. In *Conference: 2016 3rd International Conference on Modern Economic Technology and Management. Research on Modern Higher Education*, March (Vol. 2, No. 01015, pp. 1033–1042).

Weiss, W. W., Balch, R. S., & Stubbs, B. A. (2002, April). How artificial intelligence methods can forecast oil production. In *SPE/DOE Improved Oil Recovery Symposium*. OnePetro.

Wilamowski, B. M., & Irwin, J. D. (Eds.). (2011). *The Industrial Electronics Handbook: Intelligent Systems*. Boca Raton, FL: CRC Press.

Zhang, D., Yuntian, C. H. E. N., & Jin, M. E. N. G. (2018). Synthetic well logs generation via recurrent neural networks. *Petroleum Exploration and Development*, *45*(4), 629–639.

Zong, W., & Huang, G. B. (2011). Face recognition based on extreme learning machine. *Neurocomputing*, *74*(16), 2541–2551.

5 Supply Chain Management in the Oil and Gas Business

Devarani Devi Ningombam and Venkata Sravan Telu

5.1 INTRODUCTION

Managing the flow of oil and gas services between organizations and locations is referred to as supply chain management (SCM) in the oil and gas industry (Mentzeret et al., 2001). A company builds a network of suppliers to get the product from the raw material suppliers to the companies that deal directly with the customers. An organization's competitive edge increases with improved SCM (Lisitsa et al., 2019; Akinwale, 2018).

Today's supply chains are concerned with the management of data, services, and products packaged into solutions, unlike yesterday's supply chains that focused on the availability, mobility, and pricing of physical goods. Modern supply networks are curated by data scientists and analysts and leverage the vast amounts of data generated by the chain process. Future supply chain leaders and the enterprise resource planning (ERP) systems they oversee will likely focus on maximizing the utility of this data by mining it in real time with the lowest possible latency (Florescu et al., 2019). There are five components of SCM systems: planning, sourcing, manufacturing, delivering, and logistics, and returning as shown in Figure 5.1. Businesses must make smart plans to manage all resources required, balance supply and demand in order to manage inventory and the manufacturing process (Hussain, 2006). Once the supply chain is in place, choose key performance indicators (KPIs) to measure its performance in delivering customer value, efficiency, and effectiveness. This will help you avoid spending too much money on warehouses, running out of raw materials needed for production, and delays in the delivery of your products. In sourcing, we need to identify the suppliers who can provide us with the products and services we require when we need them (Sharma et al., 2022). Then, we establish procedures for managing supplier connections after that. Important procedures included in this component are ordering, receiving, controlling inventory, and approving payments to suppliers. SCM is fundamental to the overall success of a business, especially in the manufacturing industry. Manufacturers rely on their partners and vendors to keep production on track to provide them with the right type and amount of resources at

DOI: 10.1201/9781003357872-5

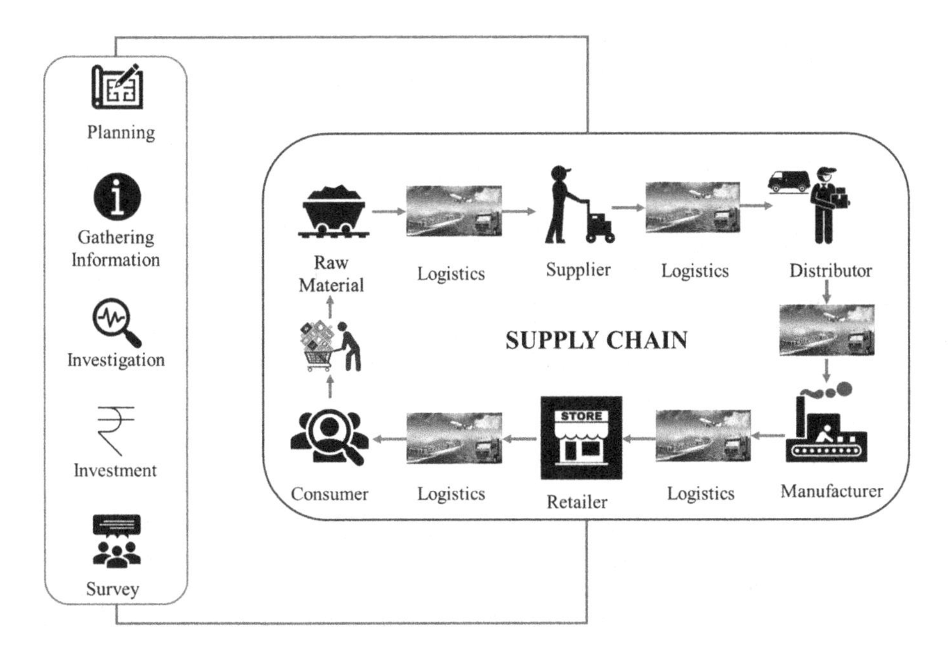

FIGURE 5.1 Framework of SCM in oil and gas.

the right time. The 2021 State of Manufacturing Report, however, states that "97% of respondents report SCM occupies a substantial amount of employee's time". If not handled effectively, SCM can be a tedious process that can inhibit yield, cause delays, affect quality, add costs, and lose money. In the manufacturing process, the raw materials we purchased are transformed into the goods that satisfy the demands of the clients. Then, we establish the processes necessary to obtain raw materials, manufacture the product, test it for quality, package it for distribution, and schedule delivery.

The next critical phase of the SCM process is the delivery of the finished product to the customer. In this process, you schedule deliveries, ship loads, invoice customers, and collect, all while coordinating customer orders. All of the preceding processes are useless if we are unable to deliver the product to the clients. Finally, reverse logistics, often known as returning goods, is a component of the post-delivery customer care process. This process establishes a system or procedure for the return of defective, extra, or unneeded goods (Hussain, 2006).

Effective SCM is influenced by a variety of elements, not just data: these include strong relationships with merchant and interim, efficient cost control, finding the right strategy lineups, and putting forefront supply chain technologies to use (Svetlana et al., 2019; Liao et al., 2015). Trade sales now undoubtedly inform the operator of stuffing orders in an industry-standard "in the nick of time" supply chain. When the merchandise is sold out, trade shelves can be restocked. One technique for improving this process is to analyze data from supply chain partners to identify areas for improvement. Three situations where effective SCM adds value to the supply chain cycle are: identifying potential problems, dynamic price optimization, and improving distribution of goods that are "available to promise". Buyers may complain about poor service if they order more goods than the manufacturer can deliver. Therefore, manufacturers might be able to predict the shortfall through data analytics before the customer is dissatisfied. The shelf life of seasonal goods is limited. These products are often discarded or offered at high prices at the end of the season (Joshi et al., 2017). Companies often dynamically adjust prices of perishable products to meet demand. Similar forecasting strategies can also increase profit margins for hard-to-perish products through the use of analytics software. According to sales forecast, original purchases, and steel production deliveries promised, analytical software tools help dynamically allocate resources and schedule operations. When the product order is placed, manufacturers can confirm the product delivery date, reducing the number of incomplete orders.

The most evident "face" of the business to customers and consumers is the supply chain. A business's SCM will defend its brand and long-term survivability better and more effectively. The key features of effective SCM are connected, collaborative, security, cognitively enabled, and comprehensive as shown in Figure 5.2. Gaining access to traditional data sets made available by traditional integration tools as well as disorganized knowledge from social media and organized knowledge from the Internet. As multi-company collaboration and engagement become more important, cloud-based trading networks are used to improve supplier collaboration. System hardening and protection against cyberattacks on the supply chain should be a major concern for the entire enterprise. By gathering, coordinating, and managing choices and actions throughout the chain, the artificial intelligence platform transforms into the control tower of the contemporary supply chain (Almuiet et al., 2019; Solano-Vanegaset et al., 2015). The majority of the supply chain is self-learning and automated. Finally, real-time data analysis needs analytics. The insights will come quickly and thoroughly.

Improved quality, lower waste and unnecessary expenses, easier configuration of distribution networks, improved collaboration, and seamless information flow are just a few advantages of an effective SCM. We will go over the key advantages of having a successful supply chain in the chapter. SCM can assist the oil and gas business cut costs, boost profitability by managing supply and coordinate delivery

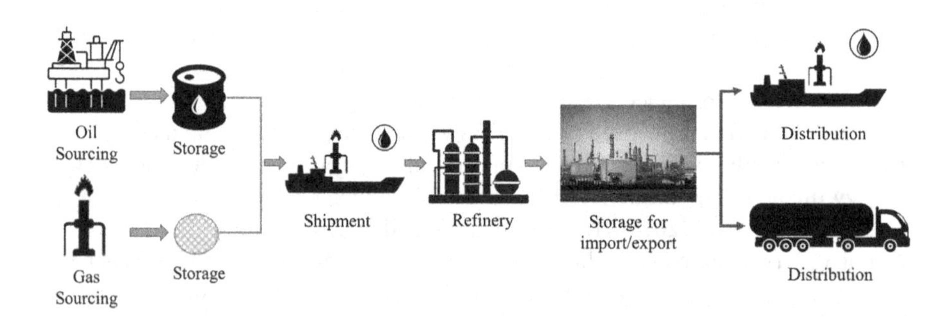

FIGURE 5.2 Key features of oil and gas business inventory value chain.

planning. Consequently, businesses can lower operating costs by utilizing technology and combining suppliers which allows them to respond quickly to client needs (Sumbal et al., 2017). Products are more likely to remain available for consumers to purchase when businesses use technology to keep closer to client demand and react more rapidly. As the global demand for oil and gas has increased, companies providing oil and gas-based products have shown volatility and been able to increase the number of customers they serve. This provides a significant impact on the development of the world economy. The rise in demand globally, as well as the supply chain's rigidity have made the management of the oil and gas business more difficult and complex.

With technological advancements, the wide network of participants involved in the various operations related to extraction and production, distribution, manufacture, storage, and consumption of oil and gas have changed. To deal with these issues, we have to analyze and understand several chain relationships. They can include inventory management, warehouse management, transportation, refining technologies to attain the customers' expectations, sales, and other processes. In addition, there are also ERP techniques that are used (Daugherty, 2011).

The purpose of this chapter is to analyze the importance of properly managing SCM in the oil and gas business to ensure that the resources are utilized efficiently. As a consequence, this in turn helps to minimize operating expenses. Abbreviations used in the study are presented in Table 5.1.

TABLE 5.1
Some Notations and their Abbreviations

Notations	Abbreviation
Supply chain management	SCM
Enterprise resource planning	ERP
Institute for supply management	ISM
Internet of Things	IoT
Artificial intelligence	AI
Warehouse management systems	WMS
Expected time of arrival	ETA
Oil-field service and equipment	OFSE
Supplier information management	SIM
Supplier relationship management	SRM
Supplier lifecycle management	SLM
Purchase orders	POs
Human-in-the-loop	HITL
Quality application development	QAD
Human resources management	HRM
Geographic information system	GIS
Enterprise geographic information system	EGIS
Return on investment	ROI

5.2 CHALLENGES IN SCM

After a very turbulent 2020, global supply networks have to some extent managed to get back on track. Still, there were some dislocations in supply chains that continued into 2021, making things difficult for both large and small companies (Baihly et al., 2010). Extreme weather occurrences, production delays, and port congestion were a few of the most frequent supply chain difficulties in 2021 (Sainath et al., 2017; Msimangiret et al., 2014). Additionally, there were numerous outliers to deal with, including the now-famous circumstance, new COVID-19 varieties, manufacturing closures, and other similar events. Many businesses managed to find innovative strategies to keep their cash flowing and their customers happy despite dealing with so many natural phenomena and external causes. While some companies moved to offering items in stock, others chose to work with domestic rather than foreign suppliers. These developments did not solve supply chain problems, but they gave brands more power and allowed them to profit in a difficult retail year (Neerosha et al., 2018).

Companies that have weathered the turmoil of the past year have likely downsized, reduced their inventories, and focused on their working capital. The supply chain crisis has undoubtedly made an impact arduous at times. Global inventory shortfalls, shipment delays and enduring backlogs of the most sought-after stock products have tested suppliers and manufacturers. For the remainder of the year, we expect supply chain and management issues to continue. These problems are typically caused by rising customer spending, a continued preference because it is more favourable to shop online and because it's necessary to make up for the delay in delivery due to various reasons.(Agarwal et al., 2016). Nevertheless, oil and gas companies are adapting to these changes by addressing inability and trying to work smarter, not harder, despite the shaky supply chain network of 2022. The key SCM challenges are shown in Figure 5.3.

The supply chain ecology is still being unset by the COVID-19 pandemic, which has brought new, unforeseen barriers to yield and profitability (Jarrahi, 2018). The top SCM issues that product-based companies from around the world will be dealing with in 2022 are listed below.

5.2.1 Inventory Management

Because of a sudden, unprecedented increase in customer demand, concerns have been raised about insufficient inputs. since the beginning of the pandemic. Despite the limited supply of oil and gas, both suppliers and retailers are still struggling to keep up with this demand. In our discussions with companies in the growth phase of the oil and gas supply network, we have identified a wide range of problems. In fact, the institute for supply management (ISM) lately assisted an audit and discovered "record long lead times, widespread shortages of key feedstocks, escalating commodity costs, and challenges with cross-industry transmission".

Digital transformation and the Internet of Things (IoT) can be a diverse grace for supply chain operations. A quantity of technological advances such as artificial

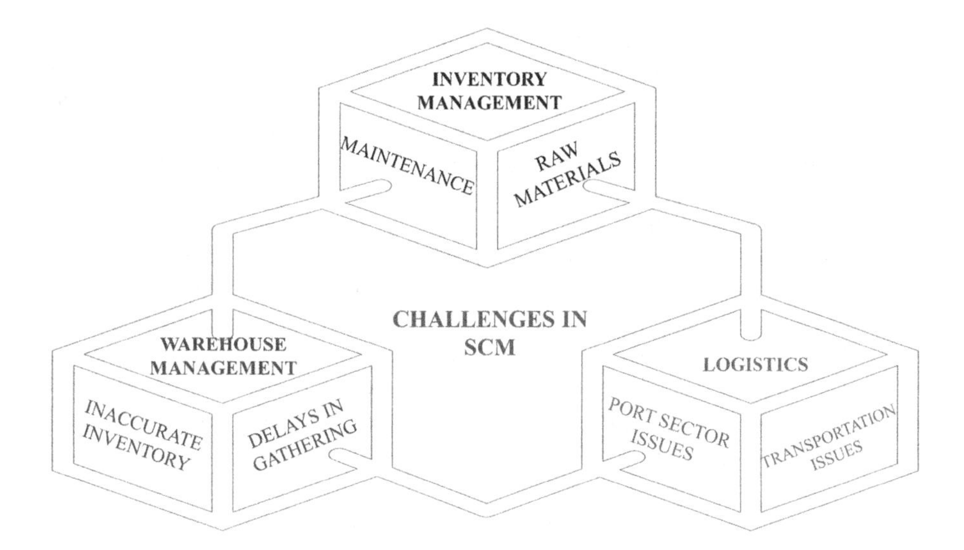

FIGURE 5.3 Challenges of SCM.

intelligence (AI), robotics and drones, electric cars, and on-demand delivery have the potential to improve the traditional supply chain (Toorajipour et al., 2021, Amirkolaii et al., 2017). The challenge is integrating these systems/services into a company's existing supply chain activities, irrespective of the fact that their long-term goal would be to boost the effectiveness and cost-effectiveness of e-commerce practices.

5.2.2 Warehouse Management

There are numerous factors driving enterprise ERP and warehouse management system (WMS) adoption, but one stands out among them all: standard data processing is ineffective and insecure. No matter how many warehouses we have, manually entering data for inventory, orders, shipping, reporting, and financing leads to chaos. Due to the lack of stock audits, even the greatest warehouse is susceptible to having inaccurate data. Exchanges and returns, for instance, could be wrongly counted, giving false information about the stock and we need to regularly check the stock (Ntabe et al., 2015).

Additionally, the pandemic has greatly changed customer perspectives and practices, lowering the bar for consignment times and raising the bar for a pleasant customer experience. The adversity is developing a malleable supply chain that leverages industrialization to maximize delivery times and easily meet increasing demand. In addition, the pandemic has driven e-commerce demand to record levels. Although a rise in desire capacity has been favourable to retailers, the main concerns for customers have been supplementary framework needs, as well as supply chain disruptions. According administration with customers over cooperative product forecasting, an averting strategy to avoid bad experiences and maintain Brad's integrity, was one of the key strategic insights (Jamaluddin et al., 2021). Authorizing these technologies necessitates some effort as well as appropriate organizational configuration, particularly when multiple warehouses are involved. However, in order to remain competitive, delivery networks must constantly evolve.

Restructuring is undeniably creating a lot of buzz among contemporary retail brands. Reshoring switching suppliers, and foreign agreements with brand new carriers are just a few of the various variations this procedure can take (Bolotina Yu et al., 2017). Choosing the correct time for a change and figuring out how to implement it as smoothly as possible are the challenges involved in restructuring. For instance, moving suppliers must be properly planned out to prevent inventory shortage throughout the changeover. In order to avoid a stockout should demand increase while waiting for replenishment or for the contracts to be finalized, we need to maintain a large amount of safety stock on hand (Jarrahi, 2018).

5.2.3 Logistics

The importance of logistics in the Indian economy is growing faster than in previous decades. In India nowadays, the logistics sector accounts for 10% of the total market. Logistics firms aid in the storage, packaging, and timely delivery of goods to customers' doorsteps as part of their present contribution to the e-commerce industry (Grant et al., 2017; Kornelyuk et al., 2018). The logistics companies in India are aware of quite a few hiccups, notwithstanding some landmark achievements. Some of the main challenges of logistics are discussed as follows:

5.2.4 Reduction of Transport Costs

One of the most important components of logistics is transportation costs. All carriers and agents try to negotiate the best prices for the transportation of each cargo (Herrera, 2012). However, this is becoming more and more difficult as fuel prices are increasing and so is the cost of transportation services, especially for imports and exports. Managers need to be well aware of upcoming purchases to optimize transportation costs, but this is challenging because customers can shop online at any time. In order to develop better strategies that maximize the capacity of containers and other objects, different modes of transportation are used. It is necessary to find the best way.

5.2.5 Processing a Large Amount of Information

While attempting to meet the supply-chain timeline, every Indian logistics company has experienced the problem of receiving a huge load of orders that would break their backs. It becomes challenging to prioritize orders and deliveries while juggling too much with too little time due to the huge volume of orders received. Due to the challenging payment periods, the problem also arises when the business has limited resources. Festival time is the best illustration of widespread supply schedule delays and a systemic breakdown of the expected time of arrival (ETA) mechanism, depending on the season.

5.2.6 Delay in Delivery

A few hours' delay in getting the proper medication to the facility could have disastrous effects, while failing to deliver high-value goods on time to the client could cost both parties lakhs of rupees. When discussing logistics, everything has a deadline. Each process is dependent on the one that comes before it, and even a small delay at one stage can have a significant impact on all the others (Melnyk et al., 2014). By the time we get to the delivery stage, the few hours of downtime have caused a delay of many days. Every logistics company has the same goal in mind: lowering the length of a supply cycle by effectively optimizing procedures. However, it is to be expected that things won't always go as planned.

5.3 OPPORTUNITIES IN THE SUPPLY CHAIN

Oil and gas supply chain businesses have seen booms and busts in recent months. Operators have drastically reduced their investments in the supply chain due to falling crude oil prices. Oil-field service and equipment (OFSE) companies are experiencing a decline in business as a result (Bresciani et al., 2016; Dickson et al., 2019). In response, they have reduced expenses and in certain cases modified their business structures. However, both operators and OFSE businesses have started collaborating in an effort to achieve long-term cost savings and profitability. In order to adapt to this changing environment, OFSE companies in particular are currently looking at five strategies: cost reduction, steep assimilation, advanced stoke models, unification, and new equipment and service models. Some of the key opportunities in the oil and gas supply chain are discussed in this section.

5.3.1 Customer Requirement Process

Every business depends on its customers. Without clients, no company in the world can continue to exist. Customers alone decide whether a company succeeds or fails. They are therefore the true leaders of a company. Understanding a customer's wants, meeting those demands, and keeping a customer is crucial for a firm. They should also focus on marketing plans to attract new clients (Koronatov et al., 2020). The only way to expand a business is to attract more and more clients. It's crucial to comprehend clients, their needs, preferences, attitudes, budgets, etc., in order to come up with creative concepts that will draw them in. Because every single customer is valuable to the company, it is crucial to recognize and categorize the different sorts of consumers and divide them into discrete categories so that effective management strategies may be developed. A business must incite a customer's appetite to purchase specific products (Unhelkar et al., 2022). To do this, the business may use a variety of strategies, including commercials, celebrity endorsements, price cuts, and free gifts, to persuade customers to purchase its goods. In the end, the customer's need determines whether or not they will buy the goods. Some of the key factors that go into a customer's decision-making before buying a product are listed as follows:

a) *Product design*: The product's design should be appealing.
b) *Functionality*: The product should include all of the features that a buyer expects when purchasing a product.
c) *Price*: The price of a product is the first item that customers consider before purchasing it. Because each buyer has their own budget, they usually stick to it unless they receive exceptional quality.
d) *Convenience*: The product and services must be convenient for the customer, or else the product will not be purchased.
e) *Compatibility*: The product should be compatible with the customer's existing products.
f) *Reliability*: The product should be dependable and suit the needs of the consumer every time.
g) *Experience*: Nowadays, everyone is busy, and they want to acquire goods that are quickly available, and there are many options available in the market for a specific product. As a result, it is critical to provide an outstanding shopping experience as well as product quality. As a result, they do not switch to a different product.

Following are the types of customers based on the consumer behaviour (Ndem et al., 2019):

i. *New customer*: A new customer is someone who has just purchased something from you for the first time. However, he purchased that product for the first time and was inexperienced with it. This is the appropriate time for you to assist them and teach them how to use the product. This allows you to build relationships with customers and convert them into loyal customers.
ii. *Loyal customer*: Loyal customers are crucial to a company's success. Not only are they loyal to the brand, but they also praise and recommend the goods to their family and friends. In this way, they also help the company with free "word of mouth" for its product. This category of customers is usually small and accounts for less than 20% of all customers, but is responsible for the majority of the company's total sales.
iii. *Need-based customers*: This type of customer only buys certain items when they need them. When they enter a store, they already know which department they want to go to. They usually do not need an assistant to select a product because they are well acquainted with the goods they want to buy. Therefore, it is important to approach them with a well thought-out strategy. Because it is very likely that they will refuse the conversation. This type of customer is easy to poach from other companies. A face-to-face interaction is the best way to approach a need-based customer. Customers who need services can be converted into loyal customers if they receive good and satisfactory service.

iv. *Impulsive customers*: An impulsive customer is a valuable asset to any organization because these customers do not shop based on necessity or a current sale. The buying habits of these customers are strongly influenced by their current mood. They are more likely to buy something if they perceive it to be useful and good at the time of purchase. Handling these customers is a demanding task since they expect immediate and succinct support with all required products available from the source. They are mostly concerned with the suggestions offered.

   A classic example of this type of customer is a woman shopping for home décor items. The most effective way to keep such customers is to offer them product discounts. This method contributes to a company's long-term profits.

v. *Discount customers*: Customers like this will never pay full price for anything. They are always seeking a good price on the item they want to buy. Such clients never purchase anything on sale. These types of consumers make up the vast majority of a company's total customers. Discount customers are the least loyal, switching when another company offers a better price.

   These clients, on the other hand, are valuable in defining the company's inventory. They typically want to know when the sale will take place and the entire details of the discount and offers made accessible by a company. When the bargain expires, these customers stop purchasing. These customers can be appeased by offering value-added services, increasing a company's chances of keeping them as a customer. They are also known as bargain hunters since they are always on the lookout for a good price.

vi. *Wandering customers*: This consumer group earns the least amount of revenue for a company. These customers have no special needs and are pulled into the store because of its atmosphere. Most of these customers like social interaction and will ask you strange questions, but they have little or no interest in acquiring products. The most common type of customer is a group of college students that visits malls to kill time. They walk into any store and ask about any random product. You should never give them too much time. However, providing them with relevant information has the potential to convert them into future customers.

In the supply chain, customer requirements are managed through the Customer Requirement Processing process component. This process includes inferring stock availability, collecting, adapting, and promoting customer demands to delivery devising and strategy, and conditioning assessment on the attainment of requirements.

Consumers are prompted to purchase according to the customer's needs for a product or service. Generally, this is driven by a number of obligations: merchandise service, cost, reliability, and simplicity. In addition, the company must manifest empathy, observance, and a successful product promotion that requires a personable approach. When these criteria are met, the company makes money, which it can use to make long-term stakes in growth and profit. Companies receive the necessary services from the Customer Requirement Processing component to successfully and efficiently handle customer needs. This process begins when the Customer Requirement Processing component receives customer requirements in the form of sales orders, customer quotations, or service orders, and it ends when these requirements are forwarded to either the Quotation and Requirements Matching process component as delivery planning requests or to the Logistics Execution Control process component as logistics execution requests. In the customer requirements processing process component, data actually moves in two directions (Shcherbanin Yu, 2016). In the direction from sales to procurement planning (with the creation of a logistics execution request and finally a delivery request), the demand is determined, the data is improved, and the customer demand is fulfilled. In the opposite direction, the sales-related process component receives the latest delivery planning and delivery data obtained from the availability check. This comprises updates to the availability confirmation as well as notifications of the requirements being met, i.e., the delivery of the goods that were first requested in the client process component, having fulfilled the requirements. As a result, businesses have realized that increased supply chain efficiency, particularly in the logistics sector, represents a significant opportunity for cost reductions. Additionally, businesses think that rather than individual businesses, the supply chain in which they engage as suppliers and customers is what drives competition.

In order for companies to meet consumer demand, the processing component communicates efficiently with the corresponding distribution, SCM, and logistics execution components. Realistic availability dates can be calculated for orders, ensuring that deadlines are met, and customers are satisfied. According to their particular requirements, this process component also gives businesses the freedom to utilize various planning methodologies, such as make-to-stock. Some of the opportunities of supply chain in customer requirement process are (Vasina, 2012):

i. *Success comes from exceeding customers' expectations*: The idea is that, especially if the objective is to exceed their expectations, it can require an inventive and proactive approach to genuinely assess client requirements. For instance, we might need to genuinely pay close attention to how they interact with a product over time. But such a strategy might yield significant benefits. According to a recent survey by The Performance Measurement Group, companies with increasing revenue are more than five times more likely to have developed systems for gathering client requirements and feedback.

   Additionally, we are aware that accurately capturing customer requirements early in the design phase can result in significant financial savings.

ii. *Discovering creative methods to hear customer voice*: We should employ a variety of techniques to effectively capture the voice of the consumer depending on the customer, the product type, the industry, and a number of other aspects. However, some strategies, such as the House of Quality, might be helpful as a starting point. The House of Quality uses a meticulously planned method and form to gather consumer requests. Another strategy is to start by getting feedback from clients about their needs, convert those needs into features, and then convert those features back into requirements.

Businesses occasionally use fictitious clients. For instance, a new product may be tested out briefly on corporate employees. The input from the fictitious clients is then gathered and examined. Contrary to actual customers, it is much easier to persuade people to complete surveys and feedback forms, especially if done anonymously. The idea of offering prototypes to seasoned users at beta sites in exchange for their comments is one that we are all familiar with. This is even more helpful if you can watch experienced users in action and listen to them talk about the product as they use it.

Sending cross-functional teams to the locations where customers utilize the product is a related and very useful tool. Here, experts from a variety of fields watch the product being used in the setting of a consumer. Additionally, employees of the fields of sales and marketing, design engineering, production, customer support, and field service may be included in cross-functional teams. A salesperson or a design engineer working alone can overlook observations or fail to understand the significance of client behaviour that the diverse members of the cross-functional team might make. These types of activities can uncover requirements or usability improvements that a list of requirements might overlook.

5.3.2 Sourcing and Supplier Management

The amount of petroleum products available to customers has a significant impact on the population's well-being as well as the economic health of the major industries, agriculture, and other sectors. The usefulness of equalizing the purchase, storehouse, and delivery of petroleum products, monitoring the process of merchandise extraction, and controlling the devising for delivery quality and transportation, given the scale of resources required to do so, is also evident from a sizeable amount of resources used, the wide range of petroleum products, and the intricacy and broadness of intelligence (Ren et al., 2010). Oil, gas, and petrochemicals require specialized means of transition similarly conduit, ships, or mercenaries, and railways because the oil industry is global. These goods are produced in a small number of places throughout the world, but demand for them is global since they serve as a vital source of energy and raw materials for several other businesses.

The practice of supplier management ensures that a company gets the most out of the money it pays to its suppliers. It is important that both the supplier and the company work together appropriately and successfully because these supplies are essential to the efficient operation of a business. A thorough supplier management policy is necessary because the firm must establish the right relationships, manage the requirements, and communicate openly with suppliers (Oksala, 2018). A corporation retains significant power if its influence is sufficient to affect the margins and output of another company. Although there are numerous oil businesses in the world, just a small number of significant companies control a large portion of the coal and oil industry. Huge capital investments have the tendency to eliminate many

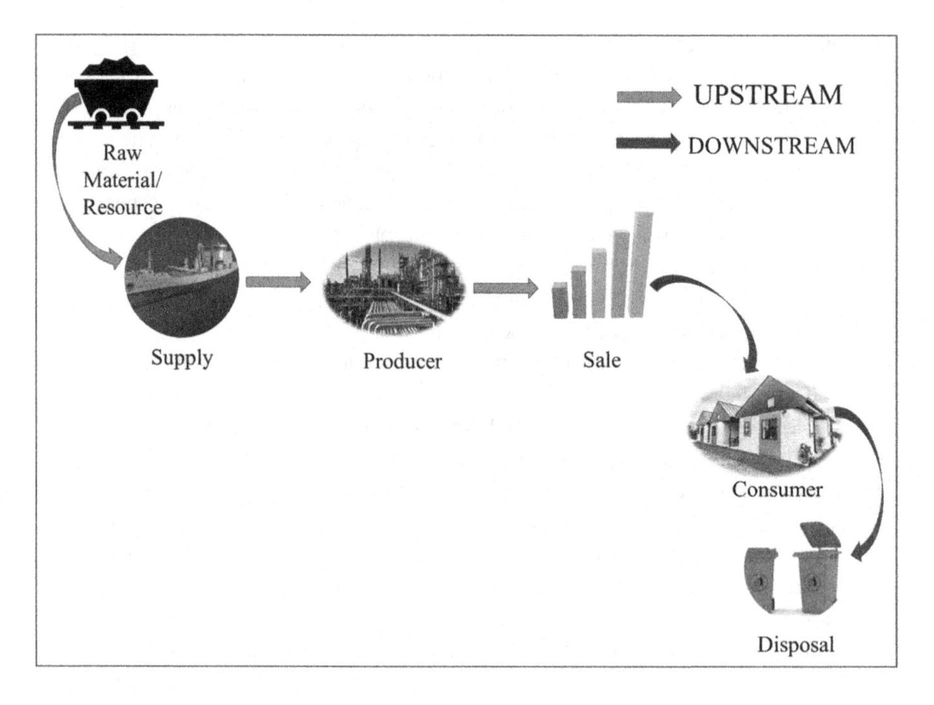

FIGURE 5.4 SCM upstream and downstream.

rig, pipeline, and refining suppliers. Although there isn't much fierce rivalry amongst them, they do have a lot of influence over smaller drilling and support businesses. As resources, a company is linked to its upstream suppliers and downstream distributors in a supply chain, information, and money flow through the chain (see Figure 5.4) (Selot, 2009).

The process by which a company collects, maintains, amends, and analyzes all stockpile knowledge in a specific region is called supplier information management (SIM). This data includes key trade information, along with the authorizations and licences needed to conduct business with the buyer (Janssen et al., 2020). This lessens the buyer's workload while improving compliance and strengthening relationships with suppliers. Supplier relationship management (SRM) is the methodical process a company uses to assess the contribution and impact of suppliers on its success, identify strategies to improve their performance, and create a strategic plan to implement the identified results.

The following is a list of the primary characters of supplier rapport management.

a. *Purchase market*: An "arm's length" partnership is usually a simple, everyday buyer/seller transaction agreement for certain goods or services. It usually involves only the fulfilment of contractual obligations; there is little to no interaction beyond the identification of the need and the fulfilment of the obligation.

b. *Ongoing partnership*: A situation where one supplier is chosen over others, either publicly or informally. Information is typically shared more frequently, and the bond between the two parties is generally strong.
c. *Partnership*: Traditionally, partnerships are a longer-term arrangement than those mentioned above. They are distinguished from continuous relationships by greater trust, extensive information sharing, and business goals.
d. *Decisive accord*: A stable association where one or more parties have settled to collaborate, commonly with some kind of limited compliance, with stated economic goals and policies. No matter how formal or causal, stimulus among the two parties is another possibility and also requires a high level of participation to comprehend promising, mostly daily coordination.
e. *Integration from the past*: In this arrangement, the company fully owns the supplier as a component of their enterprise. There is a cohesive culture as a result, and ideally, all information and plans are shared completely. Organizations and suppliers are essentially the same thing.

In contrast to SRM, which focuses primarily on managing suppliers after the contract has been signed. Supplier lifecycle management (SLM) is a holistic approach to managing high-value or strategic suppliers, from selection through relationship termination. The most crucial facet of the system is recognizing the value that suppliers can provide so that the advantages can be incorporated into acquisition mechanisms (Ganiyu et al., 2020). The supplier management process flow includes,

Step 1—Qualify: Figure out interims to regulate if they are capable of providing the essential on-set and out-set according to the buyer's criteria.

Step 2—Onboarding: The onboarding process then gathers the necessary information about the selected suppliers. Domestic practices and rules must be modified or amended appropriately, while admissible partners must be advised in order for the company to recognize and start trading with the new supplier.

Step 3—Analysis: The mechanism of allocating transitory into different transitory sections based on a pre-determined set of indicators, such as stockpile uncertainty, stockpile necessity, and overall investment, is called supplier segmentation.

Step 4—Association: Adjacent participation bounded by suppliers and vendors boosts the accord and profit-making value over practice and act growth and utility output or modernization.

Step 5—Assessment: The ultimate phase, based on a variety of indicators such as shipment time, pricing, formulation, quality, and service, is used to evaluate a stockpile achievement and safeguard that they are adhering according to the contract's terms.

The following are some of the key benefits of the supplier management process (Johnsen et al., 2008).

a. *Risk minimization*: A current supplier information management system offers the added advantage of resolving supply concerns in a way that lowers risk and also lessens the possibility of fraud, such as fraudulent invoices between personnel and suppliers.
b. *Cost minimization*: A program for managing Oil and Gas business suppliers' relationships may result in long-term cost savings.
c. *Efficiency maximization*: Oil and gas business suppliers get a better grasp of the companies they work with as the relationship develops and communication gets better. This results in fewer supply chain delays and better operational flow.
d. *Lowest possible price volatility*: Companies can benefit from scaled increases, set prices, or discounts through supplier relationship management in return for extended contract terms, minimal order amounts, and other specifications.
e. *Streamlining the supply chain in the oil and gas business*: A streamlined supply chain streamlines the purchasing process and makes budgeting easier by enabling customers to work with fewer vendors.

5.4 ANALYSIS

The intense competition in the oil and gas sector arena is forcing businesses to improve their present business models, increase automation, and maximum cost reductions across their sourcing and procurement processes. Supply chain analysis enables businesses in the oil and gas sector to give consumers top-tier goods and services at competitive prices, thus boosting their overall profitability and competitiveness. Businesses are being pushed to reduce operating expenses as a result of the recent drop in oil prices in order to maintain their position in the oil and gas sector. Businesses in the oil and gas industry have also been obliged to reduce investments as a result of the recent fluctuations in crude oil prices, switching instead to relying on supply chain analysis studies to pay fixed expenses.

Due to a significant backlog of unplaced requests, issues with historical data, and unscheduled cycled requests, a large oil and gas industry client with operations dispersed across numerous sites encountered difficulties. The client's activities also caused a rise in the number of purchase orders (POs) that were past due, increasing costs, slowing down work, and lengthening cycle times (Chima, 2007). They also experienced late payments, outdated bills, and a convoluted process for handling invoice disputes. By aligning specialized resources to reduce the time it takes to get from requisition to PO and clear the backlog in PO development, the oil and gas sector customer was able to revamp the procure to pay process during the course of this supply chain analysis project. Additionally, the customer established invoice processing procedures in line with industry norms and launched a daily goods receipt report to identify inaccurate invoices. The following fundamental inquiries are addressed in this supply chain analysis engagement:

a) What risks are often linked to the supply chain?
b) What is the present structure and status of the supply chain?

c) What possible consequences can these risks have for the oil and gas business?

a) What risks are often linked to the supply chain?

The pandemic and the temporary recession of 2020 were followed by a resurgence of the global economy. The outbreak, however, made it difficult for supply networks to keep up with the sudden increase in demand. Global "transportation problems" led to lengthy port closures, increased shipping container costs, increased overall costs, and a shocking sea-change in narratives. But the most recent supply chain risk report raises concerns about cadence. As we discuss them, don't think of each topic independently. Simply put, rather than simply one or the other, different supply chain components are increasingly being impacted by risk events. This implies that companies and supply networks are unable to recover after one setback before suffering another. Risks that are external to the user or within users' control and risks that are internal to you and within control should be covered by the risk management plan.

i. *Risks from outside the supply chain*: Events in the supply chain's upstream or downstream can act as a catalyst for external dangers (Kramer, 2022). Five categories can be used to categorize external risks: unpredictable or poorly understood consumer or end-user demand is what causes demand hazards. Interruptions in the movement of product, including raw materials and parts, throughout the supply chain, result in supply concerns. External to the supply chain issues, usually connected to economic, social, governmental, and environmental factors, such the danger of terrorism, are what cause environmental risks.
ii. *Manufacturing risks*: When we talk about manufacturing risks, we're referring to the prospect of a crucial step or component in the manufacturing practice being restrained, causing schedules to be pushed back for everyone involved. There could be a number of factors at work here, including:
   - Cost of goods and raw materials: When the price of raw materials rises, so does the price of production. As a result, businesses are frequently forced to raise their prices, which harms both their reputation and their customers.
   - Raw material scarcity: While companies are increasingly looking for local or regional manufacturing options for finished goods, many key commodities are still produced on a global scale. As a result, production bottlenecks or demand spikes are possible.
   - Distribution disruptions: Moving a major supplier's operations to another location can have a significant impact on raw material and production costs. To avoid this, risk controllers need to conduct lines of intelligence open and their agreements solid with all relevant parties.
iii. *Monetary risks*: A sudden or troublesome reversal in currency customs to a supplier's insolvency are just a few examples of the risks that could arise. Financial hazards can include going over budget, hitting a ceiling, making

progress, missing deadlines, and needing more funding, to name a few. Fiscal prospects also include abrupt charge increases that may be associated with other risk factors: variations in the outlook of the trail are desired to successfully complete the activity.

iv. *Risks involved in storage and transportation*: The mechanism of tinning and coordinating equity in warehouses centres to preserve an uninterrupted inventory of stocks in the market is called warehousing. Stored goods are vulnerable to theft, damage, and spoilage. If not taken care of properly, it can ruin the entire inventory planning process. Similar to storage, optimizing the shipping and distribution of products is also critical. Supply disruptions can have two consequences for the business: non-delivery or delayed delivery. This can happen due to transportation issues, shortfall of wagons, or other factors. This type of supply chain risk can directly clash stockpile outlay and brand prestige. Therefore, it is important to take the paramount precautions. These are the primary exposures that intimidate the agenda, but, as already quoted, they can also impact costs. They are primarily the result of poor design interpretation or poorly formulated terms of reference. Agenda changes are often cognate by logical tragedy such as monsoon, heat, or floods, or by provisional non-compliance. Scope risks can arise from changes that become necessary when the conventional Statement of Work (SOW) is no longer attainable, or from market-guided specialized revisions.

v. *Quality problems*: Quality problems in the individual links of the supply chain, e.g., raw materials, ingredients, production, logistics and packaging, lead to a domino fallout that disseminates through a multi-level supply chain. While a minor quality defect (e.g., wet and limp packaging cartons) may seem minor, the consequences can be strict (loading deceit into machines or transits, and the integrated accumulation of line comes to a halt). In case of such risks, asset liability administration provides a number of methods to find scientific explanations. Six sigma quality assurance techniques can help you automate the assembling process as much as desirable while reducing the circumstances of human error. They also build up leaner and more dynamic processes.

Supply scarcity, rising prices, and a constrained supply chain environment have enforced supply chain leaders around the world to establish solutions. Artificial intelligence (AI) is the driving force behind this innovation. Innovative companies are using AI to launch in all areas of SCM (Toorajipour et al., 2021; Bachlaus et al., 2008). AI has the potential to transform SCM practices, from pre-season requirement forecasting, wholesale accords, devising and allotment, stockpile availability, commodity selection, stockpile attainment, and itinerary escalation to last-mile practices (Avci et al., 2017). Below are some key steps companies can take to establish better AI solutions to their SCM problems:

i. *Clearly stating the issue*: A dearth of a workable hypothesis is the main cause of the failure of so many AI and machine learning efforts. Supply chain experts may not always speak the same language as AI engineers or analytics practitioners. The supply chain teams may accurately transform their key pain points into a set of problem statements by spending time with their analytics counterparts. The statement of the problem in the context of NPI forecasting would go something like this: "the organization finds it challenging to predict total sell-through demand properly". In order to reliably predict demand for the first six months following a product introduction, we wish to develop an AI model.

ii. Data engineers, data scientists, and SCM specialists need to discover institutional knowledge authorities that could help them accurately formulate their direct selling action once the problem statement is clear. This approach is most frequently used to include all potential data experts that may be useful for NPI forecasts. Teams can prioritize data sets with the highest data quality after identifying relevant data sources (Yangüez Cervantes and others, 2021). The use of data from third parties, e.g., B. competitor pricing, promotions, launches, etc., may also be required to resolve NPI and demand forecasting issues.

iii. *Modelling*: Carefully selected machine learning algorithms must be used for modelling. By transforming different data sources and developing characteristics that best account for the variability in the data, data scientists experiment with a variety of data sources. Now that AI technology has advanced, specifically in cloud computing and deep neural networks, businesses can produce cutting-edge solutions at a competitive price. As a result, companies may now utilize their resources more effectively (Al-Bulushi et al., 2009).

iv. *Model deployment and inference*: The majority of SCM tools with AI aim to contribute smooth inference. Fortuitously, considerable cloud utility provides AI and machine learning accounts that enable companies to deploy models quickly. This is not to say that AI forecasts should be used automatically; on the contrary, a number of productive initiatives employ a "human-in-the-loop" AI strategy (HITL). HITL administers a decision-making structure (Bravo et al., 2011).

Strict protocols must be in place for the managers of the oil and gas businesses to reduce the after-effects of the risks associated with the industry because it is hazardous at all stages: production, supply, transportation, distribution, and sale (Fernandes et al., 2009; You et al., 2009). Owners of businesses with high operational risks must deal with declining production output, subpar consumer goods, and inefficient production techniques. A competitor may have the opportunity to intervene in these situations and take the company's market share. In this scenario, we can derive risk as (Fletcher, 2005):

$$Risk = Consequence \times Likelihood$$

where: (i) Likelihood is the fusibility of existence of a brunt that alters the situation; and, (ii) Consequence is the effects on the environment when an event occurs.

Among all the possible consequences, revenue loss and negative cash flow are the popular consequences in the oil and gas business (Breen et al., 2012).

i. *Revenue loss*: Oil and gas business that has been in operation for a while runs the risk of losing money. If revenue falls for several consecutive fiscal quarters, a company may need to reduce operations, fire employees, or do both. If the economy is struggling and customers cannot afford to buy oil and gas, an oil and gas distributor may need to cut back and purchase less inventory. Revenue declines, on the other hand, can motivate businesses to investigate new markets or seek out new business opportunities.
ii. *Negative cash flow*: A new business often has negative cash flow when it first starts out because it spends more money than it makes. For instance, a company that borrows money to open its doors, invest in new equipment, and buy all of its initial inventory is likely to lose money in the first few quarters or fiscal year. The business might not be able to pay its debts if the cash flow is negative for more than a year. Managers must investigate cash outflows to determine whether there is fraud, waste, or any other issues that might be affecting the business. Effective budget management can prevent future negative cash flow.

5.5 INTRODUCTION TO ERP FOR SCM

SCM can develop even a significant role of an intelligent business blueprint as the operator grows. To achieve effortless and more fruitful fallouts, companies can welfare significantly from rationalizing and upgrading SCM processes through the use of ERP software. ERP is the ability to provide a coordinated set of business applications (Johnson, 2022, Lamba, 2022). ERP technologies share a common process and data model that covers a wide range of comprehensive end-to-end operational processes, including those related to SCM, manufacturing, sales, human resources and finance, among others. ERP systems collect standard data from various sources. With a single source of truth, organizations can ensure data integrity and avoid data duplication from multiple sources. That's why ERP systems are critical to running business in almost every industry.

5.5.1 Main Features of ERP

An ERP for the oil and gas sector serves multiple industries. Oil and gas production processes include drilling, extraction, purification, and distribution, and each has specific requirements for the ideal ERP system (Miller, 2022). Software developers do keep an eye out for us, and we have access to a wide range of products, some of which are tailored to a single area of the market while others are flexible enough to serve many markets. We should consider a few common criteria, and the oil and gas ERP system you select must satisfy the expectations. The ERP systems generally have the following characteristics (Jafari et al., 2018):

i. *Integration*: The ERP systems' integration sets them apart from other tools. In a modern company, working independently without assistance from other divisions is tough to imagine. Data from every department is collected, saved, and analyzed through ERP systems' enterprise-wide data integration. This has greatly facilitated the development of "a single source of truth", which reduces team differences and hence lowers possible expenses and errors.
ii. *Automation*: Order entry, payroll, accounting, invoicing, and reporting are all time-consuming tasks that can be accomplished in a fraction of the time thanks to ERP systems, freeing you up to concentrate on other crucial duties for the firm. Process automation also lowers human error.
iii. *Real-time operations*: Because all the data is kept in a single database and automation accelerates processes, it is easy for the system to react quickly to changes in orders, shipping, product availability, and other factors. Even if it isn't always real-time, it's very near.
iv. *Tracking and visibility*: Because ERP systems are so interconnected, it is feasible to keep track of everything from raw materials to finished goods as they are manufactured and distributed. A company can therefore take advantage of the enormous visibility this provides to foresee problems like a delayed time to market and stock shortages.

5.5.2 Role of ERP in SCM

The management of finances, projects, manufacturing, and operations, including natural resource management and human resource management, must be supported by an ERP system for the oil and gas sector (Paulton, 2021). The oil and gas ERP need to be able to integrate with existing operations and systems and identify supply chain issues before they become serious. You don't want a delay while resolving issues in the field or throughout the supply chain; the system must react immediately. Additionally, the demanding industry-defined compliance standards and robust reporting must be managed via the ERP system. It is impossible to emphasize the importance of ERP in SCM (Mishra et al., 2009; Romero et al., 2010).

An ERP software is directly related to corporate growth and has had a significant impact on how organizations are able to operate. For a company to grow, effective supply chain and ERP administration are essential. These tools, which are offered by supply chain specialists like quality application development (QAD), groups different supply chain operations into a single dashboard, improving visibility and simplifying interactions between vendors and suppliers (Mishra et al., 2011). Manufacturing ERP software can automate supply chain processes to increase the efficiency of workers elsewhere, as shown in Figure 5.1.

The ERPs have a variety of uses in SCM, helping with important tasks like (Jafari et al., 2018):

a. *Inventory management*: Inventory is a valuable asset that needs to be managed as effectively as possible. A business demands adequate oil and gas in a specific pattern to avoid losing any sales. But one ought to avoid

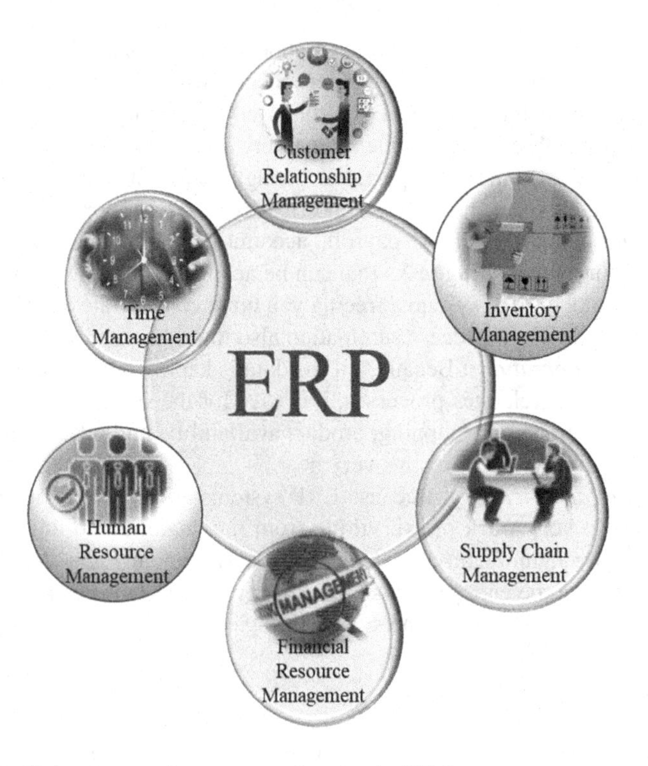

FIGURE 5.5 Role of enterprise resource planning in SCM.

spending too much. The oil and gas inventory is exceptional in that a sizeable percentage is just hazily known. Geologists' estimates that there is a specific amount underground are the basis for significant investments (Ali et al., 2020). But until the last drop of oil is emptied out, we won't know the precise amount. Even so, we routinely "find" more stock via cutting-edge techniques like fracking. Unlike many other businesses, the oil and gas industry's ERP must handle movable inventory.

The oil and gas sector needs pricey equipment. Managing billions of dollars is necessary to operate an oil refinery, a global pipeline, or an offshore drilling rig. We need the oil and gas ERP systems to help us either make sure that these assets keep producing value or decide where to use them to do so. Given the importance of depreciation and amortization as components of financial reporting, the ERP should operate effectively in this scenario to optimize profitability. Across the supply chain, ERP software can accelerate manufacturing operations by creating a bill of materials (BOM) and providing real-time data for human and mechanical resources. To minimize errors and guarantee prompt product delivery, the ERP system can also record shipment documentation. The ERP solution makes it simple to incorporate last-minute changes that are needed for specific jobs or orders.

b. *Customer*: Another area where the ERP must be helpful is in quality control, which is crucial to the oil and gas industry (O'Neal, 2014). Although the items are quite profitable, there may be environmental problems. To lessen incidents and assist with compliance reporting in refining and purification, ERP should track any distribution losses (Bae et al., 2010). The ERP must enable the conversions since distribution losses can be in the form of liquid, gaseous, or solid materials. Customers also have specific wants, and ERP should assist us in managing the business operations to satisfy every customer's expectations at the most affordable cost.

   ERP solutions have the potential to create demand as soon as orders are received, which is just one of the many automated features that make SCM processes simpler. Using ERP software, jobs are scheduled after orders are received. It is advantageous for plant managers to have access to real-time information on the resources being used to execute tasks. Additionally, it permits businesses to accurately plan product deliveries while ensuring that all of their manufacturing operations are carried out in accordance with consumer demand. When the goods are delivered to the customer, the ERP system can generate invoices that are subsequently delivered to them. The software also maintains a comprehensive record of shipment and delivery data to ensure that all orders are shipped out on time. The package quality assurance tests can be chosen and carried out by the ERP software when resolving resource disputes.

c. *SCM*: ERP programs with a focus on SCM assist in controlling the flow of resources and raw materials essential to the supply chain. ERP can be used to optimize or automate functions like managing warehouse resources, moving products, and interacting with vendors in order to speed up procedures and boost efficiency (Hayes, 2022). The supply and demand for goods and services are attempted to be profitably balanced. SCM includes the planning, execution, management, and supervision of supply activities. The objective of SCM is to foster more trust and collaboration among supply chain members. The management of the entire value creation process, from the procurement of raw materials to the delivery of finished goods to customers, is made simple by ERP.

d. *Financial resource management*: Forecasting financial data enables management to evaluate the present financial state of an organization by correlating conventional and expected monetary knowledge. Monetary outlining assembles knowledge on the prevailing budgetary year's actual facts and analyzes potential fallouts and outlooks. Using ERP safeguards that the data you collect is correct because a financial estimate is only as accurate as the data it is based on. With ERP financial management software, several operations may be simplified, accuracy may be increased, and costs may be reduced (Poston et al., 2001). Financial procedures can be automated yet analytics still offers useful insight for management choices. The ERP systems are also used to monitor business operations and prevent accounting irregularities.

Every aspect of the industry is international. Wells are located in one country while refineries are in another. Additionally, consumers are distributed around the world. Every location has a distinct tax system depending on the volume of products brought into or taken out of the region or on sales revenue. We must provide the necessary financial records to those tax and regulatory authorities for any commercial transaction. It is essential for businesses to work to minimize their tax obligations. Both the local reporting requirements and helping to balance the overall global burden must be handled by oil and gas ERP.

e. *Human resource management*: The ERP characterizes a significant part in human resource handling than what the mediocre customer could have thought up. An ERP in HRM associates with the distinct objectives descending under customer handling, like an ERP, assimilates numerous business sectors (Jamal, 2021). Labour is one of the four main resources that a corporation must have in order to produce, sell, or offer services. The focus needs to be on the employees or those who work on it while running errands for business. As a result, efforts to find the best people for a position of employment are the main focus of human resources management (HRM), also known as human asset management. Additionally, efforts are made to continuously improve the skills and abilities of those individuals by providing them with the right kind of support. If the company is large, an ERP for HRM becomes essential and inescapable.

The consolidated database is one of ERP's key advantages for HRM. A specific index makes it easier for execution to collect knowledge promptly and in an orderly manner. A HR software program allows the business to obtain intelligence about each representative, along with the name and representative code, attendance knowledge, vacation demand, work records, performance reviews, and several benefits granted, collected from the directory itself. Subsequently, it helps in administering wise decisions and provides diplomatic and well-structured output.

Talent acquisition is the second and most favourable benefit provided by an ERP. The use of ERP in HRM ensures that qualified candidates are hired for specific job openings. ERP is useful for a variety of tasks, including identifying the abilities required for a given job description, monitoring employee onboarding training, keeping track of employee performance, and more. To put it simply, an ERP at work significantly aids in controlling the skills and training of the workforce.

Thirdly, a human resource management ERP facilitates the collection of employee performance evaluations. Data retrieval serves as a company's pivot point because it demonstrates how far the enterprise has advanced in terms of its job. With the use of elements like Timesheet and Employee Appraisal, the proper data gathering on a worker's dedication and sincerity may be created. Additionally, it helps management find rapid solutions to issues whether they are difficult or delayed. In general, the ERP serves as a catalyst, enhancing internal communication and fostering mutual trust.

f. *Time management*: With the use of an ERP system, reports may be swiftly generated to demonstrate to the supplier how similar the issues are and when they first appeared historically (Zhang et al., 2022). After you have the data, we can even track the receipt of things in real time, decide how to handle lead-time fluctuations, and set realistic requirements or next steps. Additionally, it shortens lead times by streamlining the engineering and quoting processes, enabling manufacturers to plan their production based on actual data rather than educated estimates. They can reliably offer lower lead times thanks to the implementation of ERP, which boosts revenue.
g. *Control of hard and soft assets in real time*: Effective operations in the oil and gas sector depend on asset controls. Large-scale interdependence in the areas of acquisition, development, transshipment, and refinement as well as integrated search R&D and operations are to blame for this (Liu et al., 2008). The final output is defined and delivered by a constellation of industrial equipment, including heavy-duty vehicles, rigs, pipes, and more. Because of this complex set of requirements, some ERP platforms are unable to manage various components; find out if potential vendors are knowledgeable in this field.
h. *Active management of policies and compliance*: When it comes to functionality, certain ERP platforms really shine while others fall short. The federal, state, and tribal governments all have strict regulations that are applied on a quarterly basis to the oil and gas industries. These tangles of rules and policy frameworks have evolved into ineffective nests that frequently burden down oil and gas operations. However, a lot of them continue to be around forever, so ERP solutions that are more efficient than others become important measuring sticks.
i. *Uniform data processing*: The ability to use standardized methods to manage an ERP data store becomes as crucial as how rules and degrees of process efficiency apply to oil and gas operations. Standardized systems promise greater long-term stability at a lower cost, whereas the alternative situation obviously costs more over time. Therefore, this is an issue of opportunity costs in addition to the costs of upstream maintenance. However, in such a case, the core system should not be the only item that is checked; any affiliate, utility, and third-party components should also be confirmed because, if integrated, they have the capacity to either enhance or degrade the system's final performance.

5.5.3 Advantages of ERP in SCM

A properly integrated ERP system provides an information flow that is seamless between multiple departments and functions. It also helps to improve overall business operations and streamline corporate processes. Just a few key advantages of ERP are listed below:

- Reduce operating costs: With more streamlined and accurate business processes, the operating costs of a device, component, piece of equipment, or installation can be reduced to lower levels.
- Boosting production: When employees have access to the data in one place, they can organize and execute daily tasks more quickly.
- Reduce risk: Different systems are required by various departments, such as sales and finance, to perform their tasks. This might be rather risky, especially if the business is quite big. ERP systems make avoiding that much easier.
- Boost teamwork: An organization can communicate contacts, requests, and purchase orders using ERP systems. It is now simpler for them to locate the supplies they require to finish the activities at hand.

5.6 CASE STUDY

5.6.1 Geographic Information System (GIS) Solution Enhances Inventory Control

The oil and gas business is a crucial part of the economy and infrastructure of most countries in the world. It is a mainstay of the world economy because it drives progress, development, and the satisfaction of basic human needs. It also requires the most capital. The costs and income are significant. Unexpected incidents and environmental hazards also make headlines. The ever-changing dynamics of this industry motivate continuous attempts to increase efficiency and reduce risk. Recently, the oil and gas industry has adopted the geographic information system (GIS) (Abdalla, 2018); this is the sector with the greatest potential for GIS, and widespread use is increasing rapidly. In the oil and gas business, positioning of shipments and vessels establish prompt consignment of assets and maintenances, and effective feedback to emergencies. Eradicating earlier stages of the territory's knowledge of the oilfield circuition, centralized in the venture GIS, assists in the removal of framework and equity, and the reclamation of the site for use. This is where GIS and related technologies come in, with their endless applications and ongoing uses. Operations in this industry are significantly affected by geography, including resource extraction, fieldwork management, and resource transportation. A computer program called GIS collects, stores, verifies, and displays data about specific locations on the Earth's surface.). GIS can also be seen as an area that brings together several other academic disciplines. By integrating seemingly disjointed data, GIS can aid individuals and organizations in comprehending spatial relationships and patterns. It requires a technique for combining data with differing degrees of precision from various sources. The system fundamentally handles aspects of daily life; thus, it needs constant updates to keep up to date and reliable. For a sizeable fraction of the data recorded in GIS to be useful, a special method of retrieval and manipulation is needed.

The GIS system and applications deal with information that can be considered as data with unique meaning and context rather than simply facts. Through improved mapping and spatial analytics, the GIS for oil and gas boosts viable expertise conditioning and rich region intelligence. Oil and gas firms can utilize GIS to layer data for effective pipeline mapping and collect, analyze, and store information about probable drill locations. They can remotely supervise field personnel while conducting exploratory well analysis and obtain the most recent site or pipeline updates. In order to handle logistics and inventory, the corporations have to keep an eye on the movement of loaded and empty train cars. Each train car's location is frequently updated by a third-party source. A web-based GIS program to track rail carriages in real-time was sought after by the company's enterprise geographic information system (EGIS) division (Xu et al., 2013).

In order to maximize return on investment (ROI) and lower risks, organizations are beginning to recognize the importance of geospatial. Starting with the integration of geographical information into current systems, the oil and gas industry has been looking at software companies to develop software, tools, and models tailored to the industry in order to maximize return on investment. Businesses having a geographical perspective ought to take this sector into account. Operations, offices, sites, and workers are distributed geographically. To analyze and manage regionally dispersed assets and operations, such massive and complicated data is used. It is now clear that specialized maps and models are the best instruments for visualization and communication (Li et al., 2017). However, the spatial component must be handled in order to satisfy the enormous needs of this company, necessitating the use of innovative GIS solutions. Although it has scarcely been explored, geospatial technology holds immense promise for software development and employment. GIS stationing over the oil and gas field circuition is presented in Figure 5.6. The main components of the cycle are discussed below (Pandey et al., 2020):

a. Effective planning includes mapping of potential hydrocarbon accumulations, hydrological modelling, mapping of the subsurface secondary fluid migration network, mapping of downstream accumulation by DEM, etc.; geological maps, mapping of risk segments for each component of an oil field, game analysis, and terrain verification or image validation by field surveys; land appraisal, oil sublets, slabs, and companies; expedition enumeration in an implicit platform; rapid assessment and qualification of liberty using integrative asset information and assigning weights and convention; and calculation of hydrocarbon reserves or volumes, deterministic projected volumes based on grid analysis, inventory assessment and structural reasoning of good data for eccentric hydrocarbons such as shale and other techniques.
b. Seismic devising includes analysis of the topography; maps and knowledge from seismic surveys; rectification and analysis of satellite images, etc.
c. Domain actions include the use of GIS enabling inculcating around the shallow and geological constraints, increased terrain execution efficiencies for all stockpiles and valleys, symbiotic progress and work force across drill sites, vigorous risk modelling for support management, real-time asset tracking, up-to-date dated DEMs for detecting sinkholes in mining, and many other benefits.

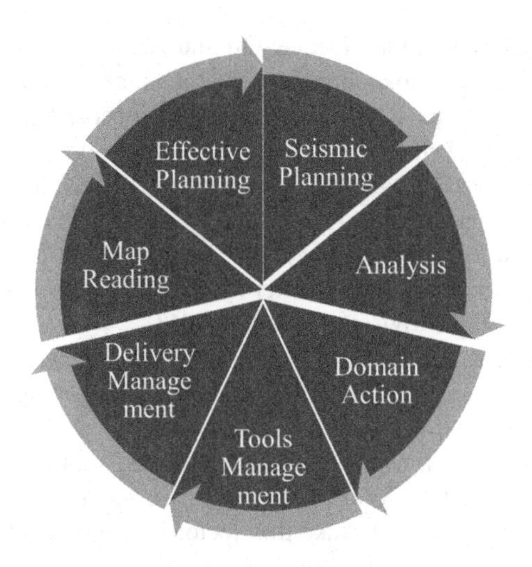

FIGURE 5.6 GIS stationing over the oil and gas field life cycle.

d. HSE (health, safety, and environment) management and tools for adversity feedback in case of an oil spill, leak, or explosion, 3D GIS with field design enables precise surveillance of relevant environmental changes in near real-time.
e. Supply management includes analysis of the most cost-effective route of the distribution network, monitoring pipelines for geohazards and flows, questing supervisions using placidly collected data, real-time surveillance and analysis of specifically distributed data, as well as seabed geodesy, asset management in offshore operations, etc., are some examples of network analysis techniques.
f. Map reading includes a GIS ecosystem that makes it easier to track significant assets in industries that require a lot of capital. Positioning of vehicles and ships guarantee efficient emergency response and timely delivery of goods and services (Evans et al., 2002).

BIBLIOGRAPHY

Agarwal, M., Sharma, R., & Mathew, L. (2016). Challenges in supply chain management in upstream sector of oil and gas industry.

Akinwale, Y. O. (2018). Empirical analysis of inbound open innovation and SMEs performance: Evidence from oil and gas industry. *South African Journal of Economic and Management Sciences*, *21*(1), 1–9.

Al-Bulushi, N., King, P. R., Blunt, M. J., & Kraaijveld, M. (2009). Development of artificial neural network models for predicting water saturation and fluid distribution. *Journal of Petroleum Science and Engineering*, *68*(3–4), 197–208. https://doi.org/10.1016/j.petrol.2009.06.017

Ali, U., Salah, B., Naeem, K., Khan, A. S., Khan, R., . . . Khan, S. (2020). Improved MRO inventory management system in oil and gas company: Increased service level and reduced average inventory investment. *Sustainability*, *12*(19), 8027. https://doi.org/10.3390/su12198027

Almuiet, M. Z., & Mohamad, M. (2019). Automated knowledge acquisition framework for supply chain management based on hybridization of case based reasoning and intelligent agent. *International Journal of Advanced Computer Science and Applications*, *10*(1), 394–403.

Amirkolaii, K. N., Baboli, A., Shahzad, M. K., & Tonadre, R. (2017). Demand forecasting for irregular demands in business aircraft spare parts supply chains by using artificial intelligence (AI). *IFAC-PapersOnLine*, *50*(1), 15221–15226.

Avci, M. G., & Selim, H. (2017). A Multi-objective, simulation-based optimization framework for supply chains with premium freights. *Expert Systems with Applications*, *67*, 95–106.

Bachlaus, M., Pandey, M. K., Mahajan, C., Shankar, R., & Tiwari, M. K. (2008). Designing an integrated multi-echelon agile supply chain network: A hybrid Taguchi-particle swarm optimization approach. *Journal of Intelligent Manufacturing*, *19*(6), 747–761.

Bae, J. K., & Kim, J. (2010). Integration of heterogeneous models to predict consumer behaviour. *Expert Systems with Applications*, *37*(3), 1821–1826.

Baihly, J. D., Altman, R. M., Malpani, R., & Luo, F. (2010, February 23–25). Shale gas production decline trend comparison over time and basins. In *Proceedings of the SPE Annual Technical Conference and Exhibition*, Pittsburgh, PA, United States. https://doi.org/10.2118/135555-MS.

Bolotina Yu, A., & Rassokha, N. S. (2017). Advantages and disadvantages of using your own warehouse in the activities of modern enterprises. *Samara State Economic University Bulletin of Young Scientists*, *1*(35), 128–130.

Bravo, C., Castro, J. A., Saputelli, L., Ríos, A., Aguilar-Martin, J., & Rivas, F. (2011). An implementation of a distributed artificial intelligence architecture to the integrated production management. *Journal of Natural Gas Science and Engineering*, *3*(6), 735–747.

Breen, P., Robinson, L. A., Rogers, S. I., Knights, A. M., Piet, G., Churilova, T., . . . Thomsen, F. (2012). The impact of risk management on the oil & gas industry. *36*(5), 1033–1043.

Bresciani, G., & Brinkman, M. (2016). *Five Strategies to Transform the Oil and Gas Supply Chain.* https://www.mckinsey.com/industries/oil-and-gas/our-insights/five-strategies-to-transform-the-oil-and-gas-supply-chain

Chima, C. M. (2007). Supply-chain management issues in the oil and gas industry. *Journal of Business and Economics Research*, *5*(6), 90–91.

Chima, C. M., & Hills, D. (2011). Supply-chain management issues in the oil and gas industry. *Journal of Business and Economics Research (JBER)*, *5*(6), 5.

Daugherty, P. J. (2011). Review of logistics and supply chain relationship literature and suggested research agenda. *International Journal of Physical Distribution and Logistics Management*, *41*(1), 16–31.

Deoliveira, J. (2008, February 25–29). GeoServer: Uniting the GeoWeb and spatial data infrastructures. In *Proceedings of the 10th International Conference for the Spatial Data Infrastructure*. St. Augustine, FL, Trinidad.

Dickson, D., Fleming, A., & Shattuck, T. (2019). *Transforming Oilfield Services.* https://www2.deloitte.com/us/en/insights/industry/oil-and-gas/oilfield-services-companies-performance-operating-models.html

Evans, F., Volz, W., Dorn, G., Frohlich, B., & Roberts, D. M. (2002, November 1). Future trends in oil and gas visualization. In *Proceedings of the Conference on Visualization 2002 (VIS 2002)*, Boston, Massachusetts, USA, October 27 (pp. 567–570).

Fernandes, L., Barbosa-Povoa, A., & Relvas, S. (2009). Risk management in petroleum supply chain.

Fletcher, W. J. (2005). The application of qualitative risk assessment methodology to prioritize issues for fisheries management. *ICES Journal of Marine Science*, *62*(8), 1576–1587.

Florescu, M. S., Ceptureanu, E. G., Cruceru, A. F., & Ceptureanu, S. I. (2019). Sustainable supply chain management strategy influence on supply chain management functions in the oil and gas distribution industry. *Energies, 12*(9), 1632.

Ganiyu, S. A., Yu, D., Xu, C., & Providence, A. M. (2020). The impact of supply chain risks and supply chain risk management strategies on enterprise performance in Ghana. *Open Journal of Business and Management, 8*(4), 1491–1507.

Grant, D. B., Wong, C. Y., & Trautrims, A. (2017). Sustainable logistics and supply chain management: Principles and practices for sustainable operations and management.

Hayes, K. (2022). ERP and supply chain management (SCM): Key differences and how they work together. https://www.selecthub.com/enterprise-resource-planning/difference-between-scm-erp/

Herrera, C. I. (2012). Outsourcing logistics in the oil and gas industry. Senior scholars thesis, Texas A&M University.

Hussain, R. (2006). Global logistics and supply chain management. *International Journal of Global Logistics & SCM, 1*(2), 90–97.

Jafari, A. A., & Nair, S. S. K. (2018). ERP implementation in the oil and gas sector: A case study in Sultanate of Oman. *7th International Conference on Reliability, Infocom Technologies and Optimization (Trends and Future Directions) (ICRITO), 2018* (pp. 848–854). https://doi.org/10.1109/ICRITO.2018.8748779

Jamal, S. (2021). Role of ERP systems in improving human resources management processes. *Review of International Geographical Education Online, 11*. https://doi.org/10.48047/rigeo.11.04.155

Jamaluddin, F., & Saibani, N. (2021). Systematic literature review of supply chain relationship approaches amongst business-to-business partners. *Sustainability, 13*(21), 11935.

Janssen, M., Brous, P., Estevez, E., Barbosa, L. S., & Janowski, T. (2020). Data governance: Organizing data for trustworthy Artificial Intelligence. *Government Information Quarterly, 37*(3), 101493.

Jarrahi, M. H. (2018). Artificial intelligence and the future of work: Human-AI symbiosis in organizational decision making. *Business Horizons, 61*(4), 577–586.

Johnsen, S., Ask, R., & Roisli, R. (2008). Reducing risk in oil and gas production operations. In E. Goetz & S. Shenoi (Eds.), *Critical Infrastructure Protection*. ICCIP 2007. IFIP International Federation for Information Processing, 253. Boston, MA: Springer.

Johnson, S. (2022). The importance of ERP in supply chain management. https://www.rootstock.com/cloud-erp-blog/importance-of-erp-in-supply-chain-management/

Joshi, P., Haghnegahdar, L., Anika, Z., & Singh, M. (2017). Supply chain innovations in the oil and gas industry.

Kleinaltenkamp, M. et al. (2014). *Business Relationship Management and Marketing: Mastering Business Markets*. Springer texts in Business and Economics.

Kornelyuk, A. M., Pavlovskaya, A. A., & Vakulich, N. A. (2018). Logistic outsourcing. Logistic outsourcing mechanism. In *Logistics-Eurasian bridge* (pp. 108–112). Krasnoyarsk, Russia: Krasnoyarsk State Agrarian University.

Koronatov, N., Ilin, I., Levina, A., & Gugutishvili, G. (2020). Requirements to IT support of oil refinery supply chain. *IOP Conference Series: Materials Science and Engineering, 1001*, 012143.

Kramer, L. (2022). Upstream vs. downstream oil and gas operations: What is the difference? https://www.investopedia.com/ask/answers/060215/what-difference-between-upstream-anddownstream-oil-and-gas-operations.asp

Lamba, I. (2022). The role and benefits of ERP in supply chain management importance of ERP in supply chain management. https://www.techjockey.com/blog/erp-in-supply-chain-management

Li, Y., Wei, B., & Wang, X. (2017). A web-based visual and analytical geographical information system for oil and gas data. *ISPRS International Journal of Geo-Information*, *6*(3), 76. https://doi.org/10.3390/ijgi6030076

Liao, Y., & Marsillac, E. (2015). External knowledge acquisition and innovation: The role of supply chain network-oriented flexibility and organisational awareness. *International Journal of Production Research*, *53*(18), 5437–5455.

Lisitsa, S., Levina, A., & Lepekhin, A. (2019). Supply-chain management in the oil industry. *E3S Web of Conferences*, *110*, 02061. https://doi.org/10.1051/e3sconf/201911002061.

Liu, S., & Xue, L. (2008, August 25–27). The application of fuzzy clustering to oil and gas evaluation. In *Proceedings of the Fuzzy Systems and Knowledge Discovery, Fifth International Conference on Fuzzy Systems and Knowledge Discovery*, Jinan, Shandong, China, 644–647. https://doi.org/10.1109/FSKD.2008.227

Melnyk, S. A., Narasimhan, R., & DeCampos, H. A. (2014). Supply chain design: Issues, challenges, frameworks and solutions. *International Journal of Production Research*, *52*(7), 1887–1896.

Mentzer, J. T., DeWitt, W., Keebler, J. S., Min, S., Nix, N. W., Smith, C. D., & Zacharia, Z. G. (2001). Defining supply chain management. *Journal of Business Logistics*, *22*(2), 1–25.

Miller, T. (2022). Essential features to look for in an oil and gas ERP system. https://www.erpfocus.com/oil-and-gas-erp-features.html.

Mishra, A., & Mishra, D. (2009). ERP system implementation: An oil and gas exploration sector perspective. *Business Information Processing*, *1*(1), 416–428.

Mishra, A., & Mishra, D. (2011). ERP project implementation: Evidence from the oil and gas sector. *Acta Polytechnica Hungarica*, 8(4), 55–75.

Msimangira, K. A., & Venkatraman, S. (2014). Supply chain management integration: Critical problems and solutions. *Operations and Supply Chain Management*, *7*(1), 23–31.

Musa, H., Nipis, V., Krishnan, P. K., Suppiah, S., & Ahmad, A. F. N. (2018). Global supply chain management: Challenges and solution. Neerosha rajah. *International Journal of Engineering and Technology*, *7*(4.34), 447–454.

Ndem, S. E., & Ebitu, E. T. (2019). Comparative analysis of business and consumer buying behavior and decisions: Opportunities and challenges in Nigeria. *British Journal of Marketing Studies*, *7*(5), 72–86.

Neerosha, R., Musa, H., Victor, N., Krishnan, P. K., Suppiah, S., & Norull Ahmad, A. F. (2018). Global supply chain management: Challenges and solution. *International Journal of Engineering and Technology*, *4*, 447–454.

Ntabe, E. N., LeBel, L., Munson, A. D., & Santa-Eulalia, L. A. (2015). A systematic literature review of the supply chain operations reference (SCOR) model application with special attention to environmental issues. *International Journal of Production Economics*, *169*, 310–332.

Oksala, M. (2018). Key factors in SME supply chain risk assessment. Master thesis, Lappeenranta University of Technology.

Pandey, Y. N., Rastogi, A., Kainkaryam, S., Bhattacharya, S., & Saputelli, L. (2020). Toward oil and gas 4.0. In *Machine Learning in the Oil and Gas Industry*. Berkeley, CA: Apress. https://doi.org/10.1007/978-1-4842-6094-4_1

Parkhi, S. (2015). A study of evolution and future of supply chain management. *AIMS International Journal of Management*, *9*, 95–106.

Paulton, K. (2021). The role of ERP in supply chain management. https://www.qad.com/blog/2021/09/the-role-of-erp-in-supply-chain-management

Poston, R., & Grabski, S. (2001). Financial impacts of enterprise resource planning implementations. *International Journal of Accounting Information Systems*, *2*(4), 271–294. https://doi.org/10.1016/S1467-0895(01)00024-0

Ren, Z. J., Cohen, M. A., Ho, T. H., & Terwiesch, C. (2010). Information sharing in a long-term supply chain relationship: The role of customer review strategy. *Operations Research*, *58*(1), 81–93.

Romero, J. A., Menon, N., Banker, R. D., & Anderson, M. (2010). ERP: Drilling for profit in the oil and gas industry. *Communications of the ACM*, *53*(7), 118–121. https://doi.org/10.1145/1785414.1785448

Sainath, L. K., & Sai, K. (2017). A study on challenges in supply chain management. *International Journal for Research & Development in Technology*, *7*, 12–16.

Selot, A. (2009). Short-term supply chain management in upstream natural gas system.

Sharma, M., Luthra, S., Joshi, S., & Kumar, A. (2022). Implementing challenges of artificial intelligence: Evidence from public manufacturing sector of an emerging economy. *Government Information Quarterly*, *39*(4), 101624.

Shcherbanin Yu, A. (2016). Logistics in the oil and gas industry: Some provisions and considerations. *Transportation and Storage of Petroleum Products and Hydrocarbons*, *4*, 22–24.

Solano-Vanegas, C., Ramos, A., & Montoya-Torres, J. (2015). Conceptual framework for agent-based modeling of customer-oriented supply networks (pp. 223–234).

Sumbal, M. S., Tsui, E., & See-to, E. W. K. (2017). Interrelationship between big data and knowledge management: An exploratory study in the oil and gas sector. *Journal of Knowledge Management*, *21*(1), 180–196. https://doi.org/10.1108/JKM-07-2016-0262

Svetlana, L., Anastasia, L., & Aleksander L. (2019). Supply chain management in oil industry. *International Science Conference SPbWOSCE-2018 "Business Technologies for Sustainable Urban Development"*, *110*, 02061.

Toorajipour, R., Sohrabpour, V., Nazarpour, A., Oghazi, P., & Fischl, M. (2021). Artificial intelligence in supply chain management: A systematic literature review. *Journal of Business Research*, *122*, 502–517.

Unhelkar, B., Joshi, S., Sharma, M., Prakash, S., Mani, A. K., & Prasad, M. (2022). Enhancing supply chain performance using RFID technology and decision support systems in the industry 4.0–A systematic literature review. *International Journal of Information Management Data Insights*, *2*(2), 100084.

Vasina, A. B. (2012). A systematic approach to stock management at oil and gas enterprises. *Russian Entrepreneurship*, 9, 85–91.

Xu, X., Shao, Y., Fu, J., Sun, Z., & Xu, X. (2013). *The Application of GIS in the Digital Oilfield Construction*. https://doi.org/10.2991/iccsee.2013.12.

Yangüez Cervantes, N., & Zapata-Jaramillo, C. M. (2021). Artificial intelligence and Industry 4.0 across the continent: How AI and 4.0 are addressed by region. In D. Burgos & J. W. Branch (Eds.), *Radical Solutions for Digital Transformation in Latin American Universities. Lecture Notes in Educational Technology* (pp. 157–177). Singapore: Springer.

You, F., Wassick, J. M., & Grossmann, I. E. (2009). Risk management for a global supply chain planning under uncertainty: Models and algorithms. *AIChE Journal*, *55*(4), 931–946. https://doi.org/10.1002/aic.11721.

Zhang, Y., Huo, B., Haney, M. H., & Kang, M. (2022). The effect of buyer digital capability advantage on supplier unethical behavior: A moderated mediation model of relationship transparency and relational capital. *International Journal of Production Economics*, *253*, 108603.

6 Prescriptive Analysis and Its Application in Oil and Gas Business

Bhagwant Singh and Monideepa Roy

6.1 INTRODUCTION

Modern analytical algorithms and advanced artificial intelligence-based models have presented the oil and gas industry with multiple opportunities to improve the overall performance of oil and gas field tasks. These technologies, such as machine learning and big data analytics, have been providing this industry with a platform to predict profitable insight from the acquired data. The processing of oil and gas data such as seismic, drilling logs, remotely sensed images, tiff files, and history data records is a huge challenge. This data must be identified, aggregated, stored, analyzed, and prefetched properly to produce valuable insight for business decision-making. Such meaningful insights can improve the effectiveness of oil and gas processing and help design advanced systems for future decision-making.

The oil and gas industry has been divided into three major sectors and providing optimized process management is the major objective of all the analytical models designed for the industry (Figure 6.1). The various tasks addressed by upstream are 1) management of seismic data produced during exploration; 2) identification of the drilling site; 3) yield of the reservoir based on the current finding and the heuristic data; 4) system failure and the process monitorization; 5) prediction of future faults; and 6) deciding course action to address the critical situations, such as blowout, fire, emission of harmful gases. Optimizing the process of extracting hydrocarbon from the reservoir requires advanced technology to manage reservoir activities via intelligent modelling. This virtual environment of the Earth's subsurface is designed with the mathematical processing of the exploration data and expert systems.

Among the various analyses that can be performed on the oil and gas data, prediction analysis is one of the most popularly used analytical models which provides the business maker with an informed opportunity to plan a strategy for extended extraction of hydrocarbon from Earth's reservoir. Such models can calibrate the system performance based on the identification of defects under extreme conditions such as environmental stress, corrosion, fatigue fractures, earthquake, tectonic displacement, etc. In the midstream, the logistics of oil and gas is another difficult area

DOI: 10.1201/9781003357872-6

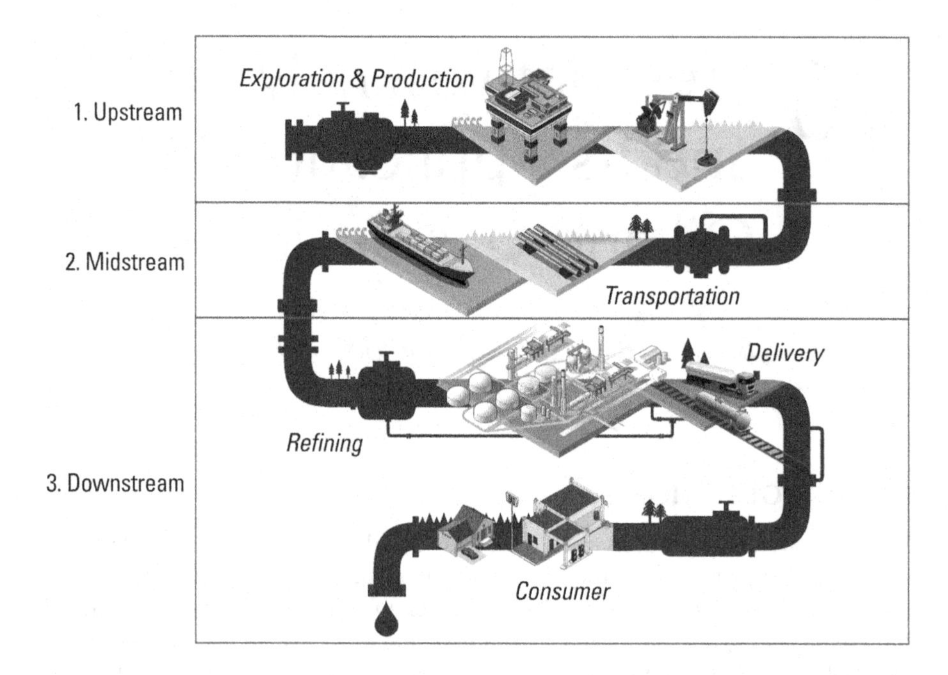

FIGURE 6.1 Workflow of the oil and gas industry. Source: https://kimray.com/training/oil-and-gas-industry-overview.

that requires advanced data analytical techniques to optimize and forecast future trends and requirements. Whereas in downstream, maintenance cost estimation, and downtime reduction are the major sectors that have adopted machine learning for improved performance of the operation.

Data representations and meta-profiling can add additional value to the strategic asset planning of future outcomes. The application of data science in the field of oil and gas has provided the business partner with various tools such as statistical modelling of the mathematical derivation, machine learning, human–machine integration, and probability identification and calculations, which can facilitate the design of the course of action of the profitable outcomes. Such tools help in integrating the data found from various resources irrespective of the data authenticity and present a well-defined and connected data format for processing. Combining historic data with real-time data required major analytical tools to foresee the path in the future.

This chapter highlights the role of analytics in the process optimization of the various stages of the oil and gas process. The chapter presents the different types of analytics that can be performed in the oil and gas industry and highlights the role of prescriptive analysis in understanding the need of the hour. This chapter introduces the major simulator used in industry from reservoir and refining simulation and justifies the application of artificial intelligence in improving the efficiency of the same.

6.2 INTRODUCTION TO DIFFERENT TYPES OF ANALYTICS

Analytics is the process of converting the raw data into meaningful insights from the data captured in the past to predict future outcomes. Every major business planner adapts these algorithms to understand the path moving forward and foresee the risk in the process execution and optimization. Four major types of analytics performed on oil and gas data are 1) descriptive; 2) diagnostic; 3) predictive; and 4) prescriptive (Figure 6.2). Each of the following analyses provides the answer to a specific question and based on the requirement any of the three can be undertaken. Descriptive is the simplest of all and prescriptive is the most sophisticated and complex task.

Descriptive analytics work on the heuristic data and present a clear picture of the data in consideration. It provides the business planner with a concrete finding of what happened and helps them in understanding the strength and weaknesses of the given data. The oil and gas life cycle includes data points from the various logs generated during the real-time processing. These data need to be studied thoroughly to understand the actual course of action undertaken in the Earth's subsurface. The study of the acquired data will help in understanding the impact of environmental and technological factors on the generated outcome. Most of the exploration sites worldwide are running at 77% efficiency (reported by McKinsey). Descriptive analytics used graphs and reports to showcase the finding of the processed data.

Secondary analytics techniques performed on the oil and gas data are diagnostic analytics. Such analytics provide a detailed answer to the reason responsible for the following outcome. The graph and figure of the descriptive-analytic can only identify the finding from the data but diagnostic analytics provide a reason for why such an event has happened. Such analytic help in identifying the future trend and establishing the relationship between the input factors and the outcomes. During the logistic management of oil and gas products, numerous factors are examined for the forecasting of the tentative maintenance plan, and hence the diagnostic analytic helps

FIGURE 6.2 Different types of analytics. Source: www.analyticsinsight.net/four-types-of-business-analytics-to-know/.

in understanding the possible reason for system failure and also present meaningful light on factors responsible for the abrupt behaviour of the machine apparatus or the design flow. These analytics also help in understanding the relationship between the market forces, the Earth's environment, and control parameters.

The third type of analytics performed in the oil and gas industry is predictive analytics. Such analysis provides the user with a clear understanding of what will happen in the future if the following pattern continues. The prediction is made on the stored data that has been captured from the various log and real-time sensed sensor data. The estimation of the yield of the reservoir before the production well is initiated is the major area that will be supported by the prediction analytics. Various environmental factors such as rock types, drilling pressure, sedimentary structure, porosity, availability of minerals, and types of fractures in the seal rock combined will produce a clear picture of the outcome of each procedure followed in the production cycle. These analytic techniques can provide the major answer to the oil and gas industry regarding what has happened in the past, the factor that was responsible for the same, and what is likely to be expected from the process undertaken for the same. All three analytics present the best outcome from the given datasets. The following sector of the oil and gas industry can be facilitated by the use of analytics algorithms such as the exploration and discovery of the new resource with the study of the seismic data and identifying the location of the traps of potential hydrocarbon.

A smart monitoring system with sensors and actuators can be designed to support the working of the production cycle and many outliers can be easily addressed via alarms. Such indications can avoid the outbreak, and identify the drilling location, and equipment health identification. The apparatus used in this industry works under physical stress and adverse environmental conditions; hence, a constant check needs to be maintained on all the instruments. An intelligent dashboard stating the health of each machinal part, the alarms designed for warning, and fault detection.

Big data analytics is the other area of application of analytics in the oil and gas sector. The global energy market has severe - disquiets about the processing and sharing of exploration data publicly, about the security prototypes (Vega-gorgojo et al., 2016). The massive data produced during hydrocarbon processing includes text files, financial ledgers, multimedia such as acoustic records, video files, remotely sensed images, sensor datasets both in real-time and cloud, tiff files, gamma logs, and calibration charts, processing such data provide the complete picture of big data in the oil and gas sector. Big data analytics has been popularized as the new trend in the execution of oil and gas data. The oil and gas industry is controlled by various organizations that provide a prototype for working with data; hence, varied data acquired justify the volume and provide a challenge to perform big data analytics. A major challenge for big data processing is the frequency and quality of the data gathered. The data in consideration showcase the ideal characteristic of big data (Perrons and Jensen, 2015) and turning Gangetic with new technological advance machinery and sensors (Feblowitz, 2013). The seismic data generated provide the volume for the processing of the oil and gas infrastructure. Pictorial and video-based datasets provide variety, logging data provide the velocity and the combination of the data from log images, and text and create messy data hence the veracity and hidden pattern and meta value of the data provide the described value to the oil and gas dataset.

The oil and gas industry includes multiple assets such as crude oil, natural gas, petrochemical distribution, storage, monitoring, retail of gas/diesel, and lubrication, therefore processing and predicting future trends need to be optimized with a high level of accuracy. Hence the most important question of how to achieve record-breaking efficiency is given by the fourth type of analysis which is prescriptive analytics.

6.3 BASICS OF PRESCRIPTIVE ANALYSIS

Statistical models designed for complex problem-solving can be facilitated by machine learning algorithms. Scanning through copious data to find a pattern and generate meaningful full insight requires enhanced algorithms to process crucial factors for efficient and quality outcomes. Identification of future action and avoiding identified bottlenecking is the objective of all advanced analytics. Prescriptive analytics is one such approach that presents the user with the pathway to solve the problem. The mathematical model that can provide the most suitable course of action considering all the possible data points, guard conditions, probability of failure, and customized weight generation is called the prescriptive analysis. It generally answers "What is required to be done to achieve the desired degree of operation efficiency?" Raw data hold much important insight and descriptive analysis helps in understanding the same, once the data is captured then the factor responsible for generated output is examined and modelled. On the bases of the raw data in the factor established the future trend of the undergoing process is designed. Now lies the most important question to improve the efficacy of the desired method and what should be done to attain the required level of system efficiency. The prescriptive analysis does the trick and hence provides a solution to multiple different industries in optimizing their process efficiency. It captures information from highly probable outcomes, and wholistic resources, and identifies results from past and current environmental parameters to generate advanced strategies to deal with the unexpected course of events. Machine learning plays an important role in designing an algorithm for prescriptive analytics. Machine-learning-based application of the argument "Well completion" has been presented by various researchers worldwide (LaFollette et al., 2014, Izadi et al., 2014, Luo et al., 2018, Guevara et al., 2018). Combing the data and interacting if and else statements can help in producing a data-driven data model that can change its answer with changing input data and hence in the oil and gas industry it plays the most crucial role in identifying the right course of action. The outcome of prescriptive analytics is based on facts and weighted projections of probability. It combines statistics and modelling techniques to produce the most optimal course of action. A large part of the research was conducted at a micro and macro level to identify the techniques that can enhance the operation of the exploration using optimized algorithms such as PSO (Shirangi and Durlofsky, 2015, 2016), GA (Onwunalu and Durlofsky, 2010), and DE (Afshari et al., 2015). Prescriptive-based analysis deployed with machine learning can address complex parameters such as "well azimuth", "drift angle", "lateral length", "fracturing fluid volume", "proppant concentration", and "coarse mesh" indicators for proppant LaFollette et al. (2014).

6.4 RESERVOIR SIMULATOR (PETREL)

Reservoir engineering can be visualized with the help of computational software called a reservoir simulator. This model combines various mathematical, geophysical, and computational models to perform prediction analytics and understand the flow of hydrocarbon from the reservoir to the production well. It is a real-time model of the process of reservoir extraction and provides a clear visualization of the various processes undertaken in the Earth's subsurface. Multiple models can be designed under any reservoir simulator such as the "geological model" to understand the geophysical property of the Earth's surface, and the "flow model" designed using mathematical derivation and heuristic data finding. This model creates the "close relationship" between the mass and the volume of fluid, and the "well model" provides the 3-D simulations of the wellbore and the environment calibration of the production process. Reservoir simulation is designed for deciding the new field development and formulating operational and financial decisions to maximize the production cycle. The processing of the data captured from the exploration site needs to be examined by the team of specialists to highlight some of the major findings such as field history, foreseeing the future outliers, and scenarios matrix.

In the process of designing reservoir simulation, experts from different fields of the oil and gas sector present the findings of their investigation. Hence, massive data is required to be captured to formulate simulation input dataset.

The main elements of a simulation study include:

- Matching field history.
- Making predictions (including a forecast based on the existing operating strategy).
- Evaluating alternative operating scenarios.

The various steps undertaken in the reservoir simulations are explained in the following diagram (Figure 6.3). The precise work of the reservoir simulator is based on the mathematical equation and model designed with the help of the research finding and historical investigation. The accuracy of these models decides the efficiency of the simulator model and presents the analysis with a clear image of the future.

The various techniques used in the reservoir simulator are Conventional Fluid Dynamics (FD) simulation provides the hypotheses of the mass conservation of hydrocarbon, the behaviour of fluid "isothermal property", and the "darcy equation" of fluid approximate through porous media such as reservoir rocks.

The thermal simulation presents the analysis with energy-efficient options in consideration of the changing environment of the reservoir, especially the temperature. A reservoir simulator undertakes input from various models such as the "geologic

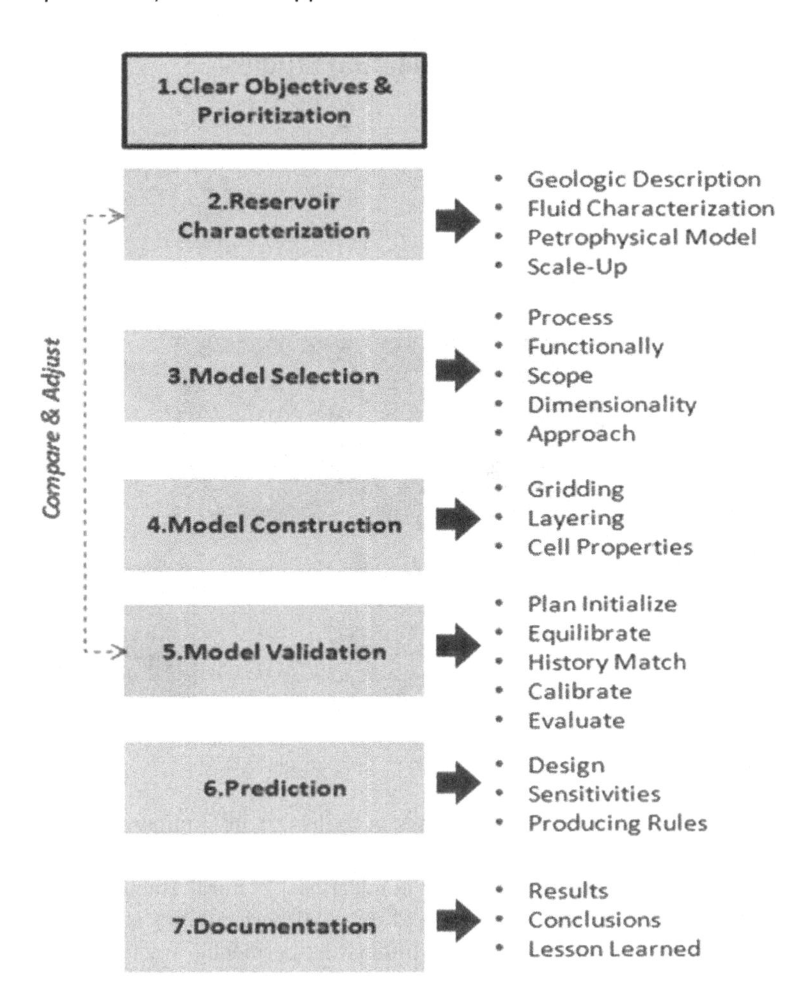

FIGURE 6.3 Flow of activities in and reservoir simulator. Source: oil and GAS portal.

model", production well data, flow matrix, and SCADA system to create a dashboard for all the activities undertaken in and around a reservoir (Figure 6.4).

The techniques used in the reservoir simulator include a "fluid dynamic simulator", which generates multiple-dimensional views of the complete exploration site or an individual well in consideration. The "cross sectional" 2-D view and "radial view" can be investigated by "the grid model". Another "finite difference model" showcases the geometry of the reservoir via a "grid line" for both the structured and the complex unstructured data points.

To find out about the flow of hydrocarbon in the well bore, running on multiple phases can be presented by "local grid refinements". Such insights help in understanding the coining of external parameters such as water or trapped gas. Refined meshing generated by the advanced algorithms assists the simulator in finding the coining outcome. Identifying faults in the reservoir and calculating the "transmissibility"

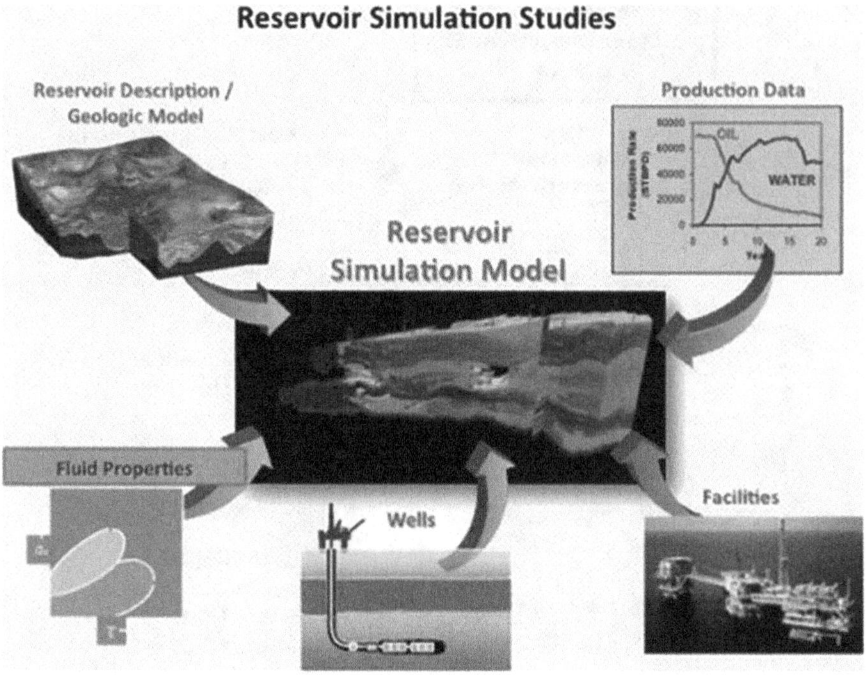

FIGURE 6.4 Reservoir simulator.

of the same are some of the other findings of the reservoir simulator. The calculation of "inter-cell flow" was computed for non-adjacent layers in the outer periphery of conventional neighbour-to-neighbour connection. "Natural fracture simulator" designed to identify the dual porosity and the dual permeability is done through a "tight matrix block" and this flow highlights the permeable fracture around the well. Another technique used in the simulator is the "black oil simulator", here the "compositional model" is created to present the PVT property of oil and gas in the form of the "equation of state" as a mixture of components. This equation can track the dynamics of various components and the phases of the hydrocarbon extractions. Simulators were designed to understand the formation of hydrocarbon in the three-phase of oil, gas, and water. The pressure at which the operation is conducted decides the execution of the "depletion study" of the reservoir. Declining pressure liberates the gas from the oil and a sudden increase in the pressure will result in an injection of water and gas in the oil. The simulator is designed to provide a clear picture of the working of the reservoir.

One of the most commonly used reservoir simulators is "Petrel". A software platform to understand the working of the exploration and production site. The software interprets the "seismic data, well correlation, calculates volume, produces a map, and suggests development strategies" to optimize the operation in the reservoir. Risk can be easily visualized by this software. The software was developed by "Technoguide" and the company was formed in 1996 by the formal employees of

"Geomatic", situated in Norway. The first version was available to the public in 1998. And in 2002 the company was acquired by Schlumberger and, since then, all the support and the new version is provided by the same company.

6.5 REFINING SIMULATOR (ASPEN—HYSYS)

Simulation is a "trial and error" testing technique used to solve complex and real-time problems. Such a system imitates the physical behaviour of the system and generates the visual next to the real system for study and investigation. The model uses mathematical calculations to predict the outcome of the calibration processes. Throughout the simulator, the user can predict the outcome of the decision taken with the help of the weighted factor and prediction model. A simple simulation can be performed in Excel but a complex process such as the fractional distillation of crude oil or the generation of petrochemical products in the refinery requires the use of advanced technologies such as machine learning and Mat-lab. The simulator is used when there is a degree of freedom at the different stages of the process and the outcome of one stage defines the income of the remaining process. The system which works on the principle of uncertainty can also be modelled using a simulator. The dependent and independent variables play an important role in designing the interaction of the deterministic and random parts of the system.

The objective of any refinery process will decide the working model of the refinery simulator. The level of complexity is decided only during the execution of the process. The refinery simulator can be modelled either for a single process which required the "first case" of complexity or for the whole cycle then the "second type" of complexity is generated where material flow, levels of the tank, movement of the fluid in the refinery (crude, by-product, and the refined product). The distillation tower is the one ideal sector that requires sophisticated techniques to address the role of artificial intelligence in simulation. Feedstocks of the polymer industry are the fundamental refinery by-products, others act as the fuels that power energy production and transportation and can be important building blocks for medicines. Automation of the various steps of the refinery can help in designing the simulator model for the whole refinery. Modern technology allows companies to use digital sensors and controls to develop incredibly accurate digital models of their assets and processes. The special process software can numerically represent the various stages of the refinery process, and moderation to these numerical values can help in understanding the complete flow of process in the refinery. Such numerical derivation can generate a realistic model of the different chemical reactions undertaken in the refinery. The data used to design the model plays the most important role in predicting the accurate visualization of the simulation process. The "parameter-based refinery" model helps in creating the gap between the model and the real world. Input from the live sensor will be accommodated in the model and help in providing a more accurate result. The refinery is the most strategically planned process in the whole oil and gas cycle. This is the location where most of the pipelines are installed from exploration to the refinery and from the refinery to the retail outlets. Monitoring the geographically distributed refinery was a major concern for most of the company in the past. With the "Internet of Things" coming into the picture, remote monitoring of various manufacturing units can be achieved. Such a model can optimize the margin of production and reduce the damage caused by environmental impact in advance. Such models help in generating multiple task execution flows to understand the effect of different parameters in controlled and uncontrolled environments.

One such chemical process simulator which used a mathematical model to understand the chemical process undergoing in a refinery is "Aspen HySYS". Calculations such as "mass balance", "energy balance", "vapour–liquid equilibrium", "heat transfer", "mass transfer", "chemical kinetics", "fractionation", and "pressure drop" can be easily modelled and monitored using it. It is used extensively in industry and academia for steady-state and dynamic simulation, process design, performance modelling, and optimization.

6.6 AI AND ITS APPLICATION IN SIMULATION

Artificial intelligence (AI) is a wide-ranging field of computer science, dealing with the design and development of intelligent computer systems. AI applications can be deployed in several areas, including data mining, natural language processing, expert systems, robotics, and machine learning. Using machine learning techniques, any software can be trained to recognize patterns in-process data and to suggest improvements to the design. This can lead to significant savings in time and money, as well as improved safety and environmental performance. By utilizing AI techniques, the software can be used to monitor process conditions and automatically adjust process setpoints to maintain optimal performance. This can lead to increased efficiency and throughput, as well as improved safety and environmental performance. Some of the important areas of the oil and gas industry which can support working are artificial intelligence are 1) "automated reservoir characterization"; AI and ML could be used to automatically characterize reservoirs, which would speed up the process and improve accuracy; 2) "automated drilling", here AI and ML could be used to automate the drilling process, which would improve efficiency and safety; 3) "predictive maintenance", here AI and ML could be used to predict when equipment is likely to need maintenance, which would improve uptime and reduce costs; and 4) "enhanced production optimization", here AI and ML could be used to optimize production, which would improve profitability. The best suitable algorithms for the optimization of the oil and gas operators are Gradient Descent, Stochastic Gradient Descent, and Adaptive Learning Rate Method.

"Property modelling" of the seismic data to generate well construction can be enhanced using machine learning applications via geostatistical models. "Sweet spot" identification and location is a major area that can be optimized by the use of such enhanced modelling. Another important application of the simulator that has been enhanced by machine learning is the classification of a given dataset. Identification of the "lithology" of the Earth's subsurface can be generated in the simulators and examined thoroughly to create a model of working and drilling of "wellbores". Such models can be presented in multi-dimensional views such as 3-D and 4-D. Python has evolved as the language of the saviour in modern times. Geoscience has witnessed the immense application of machine learning since the 1970s, but modern technology has provided the researcher with an optimized approach to capture, process, and visualize oil and gas data for predicting future outcomes. Mismanagement of the dataset has been identified as the most common error in any of the oil and gas industry situated globally and python libraries such as "sklearn", "scipy", "tensorflow", "Segyio", "Fatiando e Terra", and "PetroPy" can provide optimized solutions to multiple real-time exploration problems. An advanced feature that can be addressed by the application of artificial intelligence in the field of oil and gas is the Self-Organization of Maps (SOM). It is an unsupervised learning technique that can see past the cluttered data and detect prominent features and hence is also called a "Self-Origination Feature Map". This feature deploys "planar grid" and "codebook vector" to present "grid geometry".

With evolving technologies in artificial intelligence more refined and optimized solutions are being provided to the oil and gas industry and these can be investigated by the following case studies.

6.7 CASE STUDY ON UPSTREAM

6.7.1 Case Study 1: Optimizing the Exploration of Shale Oil

The United States have identified that reservoirs of trapped shale oil and exploring the same will put them back in the race for "energy superpower". Abundant reserves of shale oil are most difficult to locate and extract. The process of fracking has gained the advantage to extract shale oil either by "horizontal drilling" or "hydraulic fracturing". The process of fracking is the least effective and results in a severe effect on the environment around the drilling site. Hence it is mandatory to predict the finding of fracking well before the actual exploration, and big data analytics provide valuable insight into the same. Different scientific communities such as "geologists" and "geophysicists" can generate vast data from and run wells, but this is only possible when the core from the Earth's surface is extracted physically. To alleviate risk, enhanced analytical processes such as prescriptive analytics are deployed and these provide a clear picture of the drilling area and make fracking an efficient method.

Prescriptive analytics substantially reduces the financial risk of fracking by providing valuable key insights, optimal scenarios for precise drilling, and opportunity foresight. Prescriptive analytics help minimize the extent of drilling required to extract and locate shale oil. Statistics show that fracking only recovers about 20 % of the oil trapped in shale rocks. Big data analytics investigate the data captured and highlight the best site for extraction. After integrating data from distinct resources, prescriptive analytics can provide a better understanding of seismic datasets. A "depositional subsurface map" can be created using techniques such as digitization. Prescriptive modelling is the best analytical technique that can be used to make decision mapping, balancing current business demands and business objectives against constraints.

During the production phase, prescriptive analytics can predict the puncher point during the drilling process. Construction of a "wellbore" needs to be precise as it will control the risk of blowouts and common problems during hydraulic fracking. The readings such as "3-D imaging" data, logging while drilling, the value of the pressure gauge, drill bit weight, calculation of drag force, and alteration in the torque can optimize well. Prescriptive analytics can identify the time for release pressure to prevent an uncontrolled or explosive release of fracking water and hydrocarbons. Prescriptive data can not only show you where to drill, but it can help determine well productivity to assess ROI and optimize production while improving safety and mitigating risk.

6.7.2 Case 2: Permian Basin Unconventional Hydrocarbon Extraction

For the extraction of unconventional oil and gas resources, significant funds are required during drilling and completion operations. As most of the operation at the exploration site is physical and chemical models, hence a pure simulation-based optimization strategy is not feasible. Along with visualization, the data analytics approach should be applied for effective learning from historical data. The use of a predictive model can optimize the completion design of new wells. Training data include drilling, completion, and production data from previous wells in the same formation. One can employ a pipeline of model reduction and feature selection/ feature engineering methods to prepare the inputs for supervised ML algorithms. The application of data analytics and machine-learning methods to well-completion optimization has been investigated by various researchers (LaFollette et al., 2014, Izadi et al., 2014, Luo et al., 2018, Guevara et al., 2018). Papers typically investigated the application of one or two regression algorithms to predict the oil production of a well based on how the well is completed. Data cleansing and investigating data quality are major steps before applying analytics. Production, drilling, and completion data need to be cleaned. Publicly available production data has an inherent inaccuracy. Oil companies collect and maintain per-well production data, typically obtained from flowmeters. This proprietary data is more accurate. Though in both cases, there are missing data points that must be inferred. A data analytics workflow is developed and applied for finding an optimal completion strategy. Descriptive analytics is first applied to investigate and assess data quality issues. A method is presented to detect and impute bad-quality production data. Then, predictive analytics through an "AutoML" framework generates hundreds or thousands of models to find the most accurate model. In "AutoML", the use of various parametric models such as support vector regression, as well as non-parametric models including decision trees and random forests are compared. Finally, prescriptive analytics are applied using an optimization algorithm.

Data description: Data from more than 100 horizontal multi-fractured wells in the Permian Basin provided by Diamondback Energy. These wells span four different producing formations – Lower Spraberry, Middle Spraberry, Wolfcamp A, and Wolfcamp B. The dataset consists of both customer-provided and public-domain data. Directional surveys, daily production data, artificial lift method, formation tops, and stage-level completion data such as the number of clusters per stage, ISIP, proppant amount, grain size and type of proppant, fluid volume and type, and stage length were provided by the customer. In instances where some features from customer data were missing, public data was utilized.

Machine-learning model: After a rigorous feature selection process, more than 100 features were winnowed down to 15. The final list of features includes the artificial lift method, and completion-related parameters such as the number of clusters per stage, stage length, amount of injected proppant, and fluid volume normalized by lateral length. Wellbore placement-related features such as surface latitude and longitude, undulation index, toe-up/down, landing zone, and least distance to the nearest well in both the horizontal and vertical planes also made it to the final list. Using AutoML, a "gradient-boosting" regression model was selected for optimized results. The time input (the month of production) appears to be the most important feature in determining the nth month of oil production. As expected, completion features such as the amount of proppant and fluid injected, proppant concentration and clusters per stage appear high on the relative importance scale. The toe-up/down configuration of a well can have a significant impact on production. In this study, we see that on average, toe-up wells have higher production compared to toe-down wells. In unconventional basins, operators typically drill a minimum number of wells required to hold the acreage during the initial phases of the development of a play. These wells are called "parent" wells. Thereafter, as the play becomes more mature, infill well development commences (Lindsay et al., 2018). The performance of these infill or "child" wells depends on the degree of pressure depletion near parent wells and the resulting fracture interference between parent and child wells. Usually, this interference has a detrimental impact on production.

Therefore, capturing this parent–child relationship is essential for the accurate prediction of good performance. In this study, this effect was quantified using features such as the number of parent wells each well has, and features relating to the least distance to the nearest well in both the horizontal and vertical planes. However, after the feature selection process, only the latter two features figured prominently in the importance list and were selected. These last distance-related features were calculated by considering only the wells that existed before the well of interest was drilled. If new wells come online during the first 12 months of production of the well of interest, then those wells are also considered in the least distance calculations. Therefore, both minimum XY distance and minimum vert. distance are functions of time, and act as proxies for the parent–child relationship in the model.

The degree of undulation in the wellbore can also affect production. At times, there could be zones of liquid loading in areas of high undulation in the wellbore, which could in turn be detrimental to production (Yalavarthi et al., 2013). In the current study, the degree of undulation in the wellbore is quantified in a feature called the undulation index.

Hydraulic fracture height growth can vary significantly, depending on the landing point within the reservoir, due to variations of rock mechanical properties and stress profile in the reservoir (Pankaj et al., 2018). The choice of the appropriate fracturing fluid type also depends on the fracture height, which in turn is a function of the landing zone. Hence, identifying the right landing zone in a formation is crucial. In the present study, the landing zone is one of the most important features affecting production. For the Lower Spraberry formation, observations show that the optimal landing zone is closer to the bottom of the reservoir than to the top. A potential reason could be the presence of a water-bearing zone above.

Rock property data such as porosity and permeability are not usually readily available, which makes it challenging to use those in this data analytics framework. However, as we aim to develop a general prescriptive analytics framework based on data that is typically available on a large set of wells, we used a combination of surface latitude and longitude along with formation as a proxy for geology in the model.

Optimization results: After finding the optimized location of the most optimal wells.

6.8 CASE STUDY ON DOWNSTREAM

6.8.1 Case Study 1: Supply Chain 4.0

There are many gaps in the supply chain which have been recently brought to light by market turbulence and changes in the business world. Supply chain management in Industry 4.0 includes production, procurement, warehousing, inventory management, and logistics. Supply chain 4.0 is all about the application of the "Internet of Things", "robotics", "big-data", and "predictive analytics" in supply chain management. Smart manufacturing processes and workflows generate extremely large volumes of data, but the vast majority of it is useless without ML models and AI workflows to identify, infer, or act upon patterns in the data. Advanced analytics solutions support many IoT use cases, including predictive maintenance and intelligent workload optimization.

6.8.2 Supply Chain and Predictive Analytics:

Client: Global Fortune 100 multinational engineering and technology company based in Germany. The company operates through a complex network of over 440 subsidiaries and regional entities in over 60 countries worldwide.

Challenge: This company has factories, warehouses, and suppliers in various geographic locations. The major challenge is managing the circulation of raw materials and finished goods among them. A logistics platform was deployed but failed to capture all the key points of the process, hence there was a need for an update.

Solution: "N-iX" provides an advanced logistics platform catering to the advantage of predictive analysis and machine learning. Some of the features of the proposed platform are 1) "Microservices migration": a new cloud-native infrastructure of the platform based on "Azure Kubernetes", along with the suggested tech stack and the most efficient roadmap; 2) "DevOps": build a "DevOps" pipeline from scratch, setting up the environment for development and QA in "Azure", and introducing CI/CD processes that allow us to easily assemble and deploy microservices to the environment; 3) "Computer Vision": enhancement of docks using CV allows contactless tracking of goods with industrial optic sensors and "Nvidia Jetson" devices. Continuous Delivery for Machine Learning was introduced in the CV algorithm for fast execution, it allows for recursive cycles of training, testing, deploying, monitoring, and operating the ML models; and 4) "Multiplatform CV" mobile app: it covers object detection, package damage detection, OCR, and NLP for document processing.

6.8.3 Supply Chain and Big Data Analytics:

Client: A Fortune 500 industrial supply company that offers over 1.6 million quality in-stock products in various categories such as safety, material handling, and metalworking. The company also provides inventory management and technical support to more than 3 million customers in North America.

Challenge: Effective management of large amounts of data, including data on inventory-related costs. Significant overhead and operational costs were caused by on-site consultants.

Solution: Unified "cloud-based" big data platform for effective inventory management. "Snowflake" and "Airflow" technologies help in automating the data extraction process. Also, "Snowflake" minimizes data doubling by checking whether the acquired files have been already processed or not. "N-iX" managed to integrate more than 100 different data sources into a unified data platform. This includes daily data loads, along with a backfill of historical data. To migrate from the on-premise "Hadoop Hortonworks" cluster to "Amazon Web Service" and allow the processing of additional data in AWS, the N-iX team built an AWS-based big data platform from scratch. "Teradata" is used to collect data from other systems and further generate reports with "Business Object" and "Tableau". The data sources are "MS SQL", "Oracle", and "SAP". "Snowflake" architecture was used to meet the client's approach of "cloud neutrality" and it can quickly scale up and down any amount of computing power for any number of workloads and across any combination of clouds.

6.8.4 Case Study: The Norwegian Continental Shelf

Norwegian Continental Shelf (NCS) is the optimal choice to study the use of big data analytics in the upstream oil and gas industry. The NCS is rich in hydrocarbons that were first discovered in 1969, while commercial production started in the "Ekofisk" field in 1971

Petroleum activities in Norway are divided into policy, regulatory and commercial functions: Norway's policy orientation is focused on maintaining control over the oil sector; the Norwegian Petroleum Directorate (NPD) (Norvegia petroleum directorate, 2016) is the regulator body, while petroleum operators compete for oil through a licensing system. Overall, this separation of concerns is considered the canonical model of good bureaucratic design for a hydrocarbon sector (Turber et al., 2010).

Petroleum activities in the NCS have prioritized long-term R&D and tackled technically ambitious projects (Turber et al., 2010). This focus on technology makes the NCS especially attractive for a big data case study.

Participation organizations: Various organizations like Statoil, Conoco, Phillips, Lundin, Eni Norge, SUPPLIER, and NPD participated in this study (Curry et al., 2014). The impact of information technology (IT) in the industry is an evaluation of an oil and gas organization's current and future information system's needs (Nolon et al., 2005) can be classified into four roles. See Table 6.1.

Data collection activities: To gather evidence about big data uptake and the societal impacts of big data in the oil and gas industry, various interviews were conducted. The research agenda included big data uptake and big data societal impacts. Multiple sources of evidence were employed to augment the validity of our findings, as recommended in Creswell et al., 2009 and Yin et al., 2014. The model adopted a semi-structured interview approach letting the interviewees speak and elicit their views and opinions, while we aimed to cover all the topics in the agenda and to request further explanations. After each interview, the transcript was prepared, and internally revised, and then the report was shared with the interviewee (normally within a week), receiving some minor amendments in two cases. Single interviews of approximately 80 minutes were held with [I-LU], [I-ENI], and [I-SUP]. [I-ST] interview was conducted four times, [I-CP] twice, and [I-NPD-1] and [I-NPD-2] were both interviewed on two occasions at the same time. In these latter cases, the sessions were shorter (ranging from 30 to 60 minutes), but we had more time available in total. Overall, 11 interviews were conducted for this case study from December 2014 to April 2015. Besides the interviews, a workshop on big data in oil and gas (Byte et al., 2016) was held in April 2015. The program included an invited talk from

TABLE 6.1
Role of Information Technology in Oil and Gas Industry

Support role (defensive)	Current IT is not mission-critical for business operations, and new IT will offer little strategic differentiation
Factory role (defensive)	Current IT is critical for the organization's business, but new IT will offer little strategic differentiation
Turnaround role (offensive)	Current IT is not mission-critical for business operations, but new IT will be fundamental for the future
Strategic role (offensive)	Current IT is critical for the organization's business, and new IT will be fundamental for the future

the oil company Statoil, another one from the supplier National Oilwell Varco, a preliminary debriefing of the case study results, and a focus group session.

Outcomes: Concerning the data analysis, a procedure-based approach is applied to the qualitative (Creswell et al., 2009) and case study (Yin et al., 2014) research literature. In the first step, the transcripts of the interviews were prepared. Next, we read through all the transcripts and began the coding of the data. This is the process of organizing the material.

BIBLIOGRAPHY

Afshari, S., Aminshahidy, B., & Pishvaie, M. R. (2015). Well placement optimization using differential evolution algorithm. *Iranian Journal of Chemistry and Chemical Engineering (IJCCE)*, 34(2), 109–116.

British Petroleum Company. (2017). *BP Statistical Review of World Energy*. London: British Petroleum Co.

BYTE Project. (2015). Big data in oil & gas workshop. http://byte-project.eu/wp-content/uploads/2015/05/BYTE-Oslo-events-agenda-April-2015.pdf; last accessed March 2016.

Creswell, J. W. (2009). *Research Design: Qualitative, Quantitative, and Mixed Methods Approaches*, 3rd edition. Sage Publications, Uttarakhand.

Curry, E., Freitas, A., Vega-Gorgojo, G., Bigagli, L., Løvoll, G., & Finn, R. (2014). Stakeholder taxonomy. Deliverable 8.1, *BYTE Project*.

Feblowitz, J. (2013). Analytics in oil and gas: The big deal about big data. *SPE Digital Energy Conference*. Society of Petroleum Engineers. Texas, USA

Gharib Shirangi, M., Oruganti, Y., Wilson, T., Furlong, E., Winter, E., Martin, J., & Yancy, R. (2019). Prescriptive analytics for completion optimization in unconventional resources. SPE Western Regional Meeting. https://doi.org/10.2118/195311-ms

Groulx, B., Gouveia, J., & Chenery, D. (2017). Multivariate analysis using advanced probabilistic techniques for completion design optimization. *Proceedings of the SPE Unconventional Resources Conference*, Calgary, Alberta, Canada.

Guevara, J., Kormaksson, M., Zadrozny, B., Lu, L., Tolle, J., Croft, T., Wu, M., Limbeck, J., Hohl, D. (2017). *A Data-Driven Workflow for Predicting Horizontal Well Production using Vertical Well Logs*. arXiv preprint arXiv:1705.06556.

Guevara, J., Zadrozny, B., Buoro, A., Lu, L., et al. (2018). A hybrid data-driven and knowledge-driven methodology for estimating the effect of completion parameters on the cumulative production of horizontal wells. *Proceedings of the SPE Annual Technical Conference and Exhibition*, Dallas, Texas, USA.

Izadi, G., Junca, J. P., Cade, R., & Rowan, T. (2014). Production performance in Marcellus Shale: Multidisciplinary study of hydraulic fracturing. *Proceedings of the Abu Dhabi International Petroleum Exhibition and Conference*. Abu Dhabi,UAE.

Kottasová, I. (2017, September 19). This pension fund is now worth $1trillion. Retrieved July 4, 2018, from http://money.cnn.com/2017/09/19/investing/norway-pension-fund-trilliondollars/index.html

LaFollette, R. F., Izadi, G., & Zhong, M. (2014). Application of multivariate statistical modelling and geographic information systems pattern-recognition analysis to production results in the eagle ford formation of South Texas. *Proceedings of the SPE Hydraulic Fracturing Technology Conference*, The Woodlands, Texas, USA.

Lindsay, G., Miller, G., Xu, T., et al. (2018). Production performance of infill horizontal wells vs. pre-existing wells in the major US unconventional basins. *Conference Proceedings of SPE Hydraulic Fracturing Technology Conference & Exhibition*, The Woodlands, USA.

Luo, G., Tian, Y., Bychina, M., & Ehlig-Economides, C. (2018). Production optimization using machine learning in Bakken Shale. *Proceedings of Unconventional Resources Technology Conference*, Houston, Texas, USA.

Nolan, R., & McFarlan, F. W. (2005). Information technology and the board of directors. *Harvard Business Review*, 83(10), 96–106.

Norwegian Petroleum Directorate. (2016). *Norwegian Petroleum Directorate Website*. http://npd.no/en/ Last accessed March 2016.

Onwunalu, J. E., & Durlofsky, L. J. (2010). Application of a particle swarm optimization algorithm for determining optimum well location and type. *Computational Geosciences*, 14(1), 183–98.

Pankaj, P., Shukla, P., Yuan, G., & Zhang, X. (2018). Evaluating the impact of lateral landing, wellbore trajectory and hydraulic fractures to determine unconventional reservoir productivity. *Conference Proceedings of the 80th SPE EAGE Conference and Exhibition*, Copenhagen, Denmark.

Perrons, R. K., & Jensen, J. W. (2015). Data as an asset: What the oil and gas sector can learn from other industries about "Big Data". *Energy Policy*, 81, 117–121.

Shirangi, M. G., & Durlofsky, L. J. (2015). Closed-loop field development under uncertainty by use of optimization with sample validation. *SPE Journal*, 20(5), 908–922.

Shirangi, M. G., & Durlofsky, L. J. (2016). A general method to select representative models for decision making and optimization under uncertainty. *Computers & Geosciences*, 96, 109–123.

Shirangi, M. G., Volkov, O., & Durlofsky, L. J. (2018). Joint optimization of economic project life and well controls. *SPE Journal*, 23(2), 482–497.

Syaifullah, K. (2019). The use of big data in the oil and gas upstream industry. *Indonesian Journal of Energy*, 2, 14–28. https://doi.org/10.33116/ije.v2i1.31

Thurber, M. C., & Istad, B. T. (2010). *Norway's Evolving Champion: Statoil and the Politics of State Enter-prise*. Program on Energy and Sustainable Development 92, Stanford University.

Tormodsgard, Y. (2014). Facts 2014. The Norwegian petroleum sector. Technical report, *Norwegian Ministry of Petroleum and Energy and Norwegian Petroleum Directorate*.

Vega-gorgojo, G., Fjellheim, R., Roman, D., Akerkar, R., & Waaler, A. (2016). Big data in the oil & gas upstream industry – A case study on the Norwegian continental shelf. *Oil Gas European Magazine*, 67–77.

Yalavarthi, R., Jayakumar, R., Nyaaba, C., & Rai, R. (2013). Impact of completion design on unconventional horizontal well performance. *Proceedings of Unconventional Resources Technology Conference (URTeC: 1558757)*, Denver, Colorado, USA.

Yin, R. K. (2014). *Case Study Research*, 5th edition. Sage Publications, Thousand Oaks, CA.

7 Future Challenges in Petroleum Sector and IT Solutions

Kingshuk Srivastava and Madhu Khurana

7.1 INTRODUCTION

The hydrocarbon industry has always endured problems in supply activity and price. The crude oil and natural gas (O&G) industry is not new to supply disruptions and price instability. One can understand that the volatility in the pricing of crude has seen a peak at $100/bbl in 2014 and a low of $37/bbl in 2020 (Allan et al., (2007)); any industry is highly impacted by this type of high variation in pricing. Incidentally, the different factors impacting these problems are drastically different from what has happened before, and today one has to consider the combined effect of geopolitics, economics, trade policy, and different financial considerations which lead to under-investment and readjustment in the market. The three different components which need to be balanced are security, diversification, and transition of energy; this is depicted in Figure 7.1.

In the following part of this chapter, the different applications of the data and the importance of it is described with the different integration challenges it is facing in the contemporary world.

7.2 CHALLENGES IN THE OIL AND GAS INDUSTRY

The most important and vivid challenges in the industry today can be broadly differentiated into three categories. Primarily, the biggest challenge is to produce more energy at a lesser cost and lower emissions to take environmental factors into consideration. With the growing population the demand for energy is also increasing alongside the demand for cleaner energy to reduce global warming. There is also competition in this field of green and sustainable energy which is also coming of age in the context of technology, and the projection to achieve net zero emissions by 2050. The product price is mostly out of control from the company's hand, but the production process cost has been drastically reduced through the application of different IT technologies such as prediction modelling and simulation.

The three main challenges are enumerated below:

DOI: 10.1201/9781003357872-7

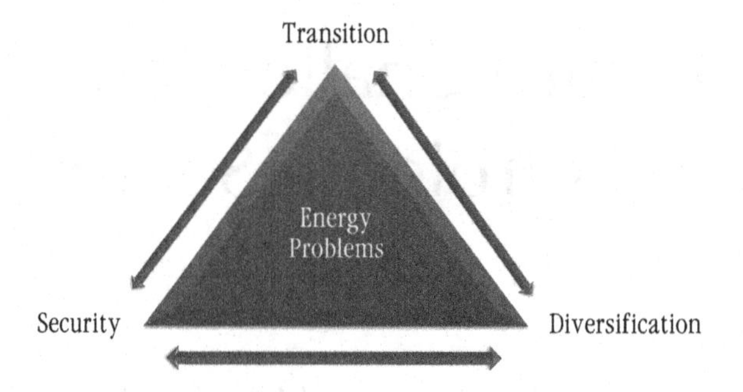

FIGURE 7.1 Energy problems. Source: Amy Chronis et al., Striking the balance: How and where will oil and gas producers deploy their cash? Deloitte Insights, August 25, 2022.

7.2.1 Reduce Cost of Production to Remain Competitive in the Market

The biggest challenge which is motivating different research in the field of oil and gas is the different ways of reducing cost of production to be able to remain relevant and competitive in the market. The exploration cost is very high, and although there has been a lot of technological progress made in avenues such as "Image Processing", "Remote Sensing", "Simulation Systems", they remain highly expensive, so the main area where a lot of optimizations are made is extraction and refining of hydrocarbons, thus managing the higher cost incurred in exploration. But there are a lot of areas where with the implementation of Artificial Intelligence and automation could further augment the cost.

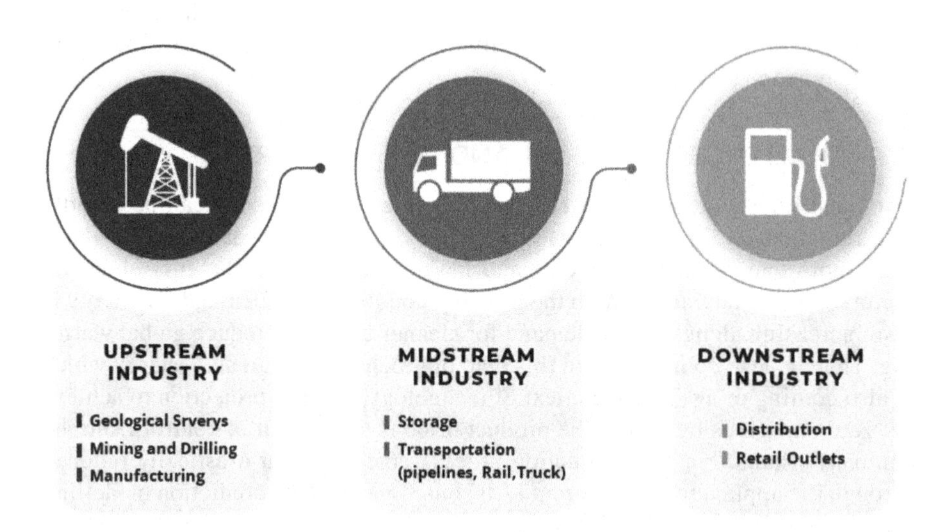

FIGURE 7.2 Sectors in the hydrocarbon value chain.

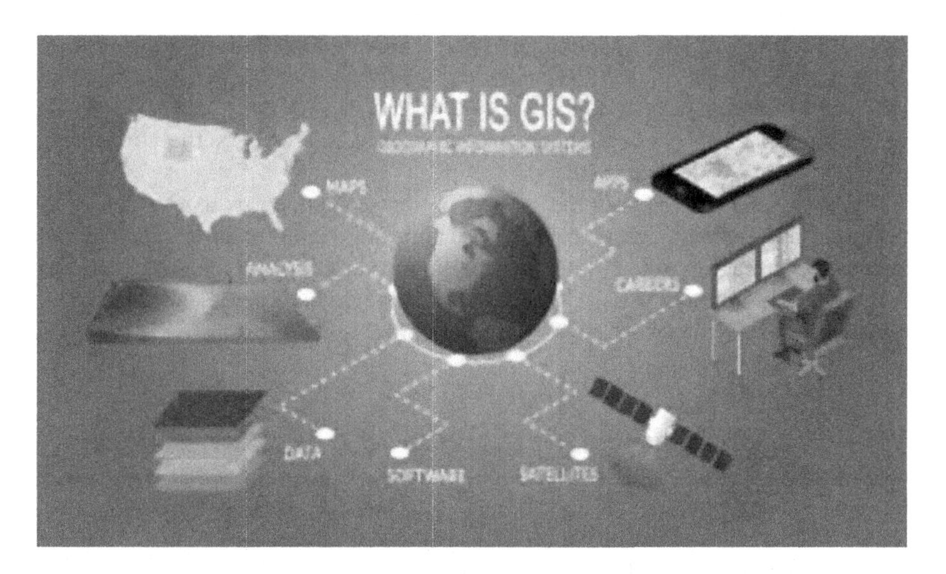

FIGURE 7.3 GIS components.

7.2.2 Improving Performance to Ensure the Valourization of Assets

One of the biggest challenges is to maintain and sustain the production quantity and withstand the fall of production at the maturing sites. To maintain the supply, new reservoirs need to be explored and developed on time, and maturing sites need to be optimized. Also, to achieve the proper supply of oil and gas products one needs to be able to maintain the reliability of their plants (no unplanned downtime, secure industrial assets, and increased throughput).

7.2.3 Reducing Carbon Footprint to Meet Stringent Governmental Standards

The oil and gas industry lies at the top of the list of industries which cause the biggest environmental impact, such as burning flue gases, proper disposal of toxic waste and by-products, and the biggest consumer of fresh water. With so much at stake, governments around the world have become stricter in the matter of implementing the environmental norms. The companies also have to guarantee and maintain transparency in the environmental management of their activities.

7.3 DATA AS A NEW OIL

> Data is the new oil. It's valuable, but if unrefined it cannot really be used. It has to be changed into gas, plastic, chemicals, etc. to create a valuable entity that drives profitable activity; so must data be broken down, analyzed for it to have value.
>
> **—Clive Humby, 2006.**

A British mathematician and data scientist Clive Humby coined the term "Data is the new Oil", in 2006. Again in 2011 senior vice president of Gartner Peter Sondergaard built on this concept to give another tag line:

"Information is the oil of the 21st century, and analytics is the combustion engine"—Peter Sondergaard, 2011.

In the oil and gas industry the most important component is data which gives the most important foresight into market dynamics through predictive analytics and automation through Artificial Intelligence. But the handling and integration of data is a big challenge, and, if not aligned with a strategic plan, may lead to a bigger disaster. Below, some of the most prominent processes are discussed to enable a robust data-handling approach.

7.3.1 Data Refining

One of the parameters which is common between oil and data is that both need a lot of processing before they can be of any use. The oil needs to be refined to make usable products such as petrol and diesel, just as data needs pre-processing to remove discrepancies and flaws before it can be analyzed.

Some of the prominent problems in real-world data are as follows:

- Inconsistent or inaccurate data.
- Missing data.
- Misrepresentation of data samples.
- Not in proper format.

Just raw data on its own is not very useful for any constructive decision-making. The data needs to be reliable, accurate, and properly represent the market. Even then, the data needs to be converted into a proper format for the purpose of analysis.

7.3.2 Data Quality

As Sondergaard had mentioned that "data is the new oil and analytics is the combustion engine". Contemporary "Artificial Intelligence" demands a colossal quantity of data to be of any usage in modern AI systems. For any type of automation processes, a lot of temporal data are required to train the different required models.

The realization of the value of data which is being generated everyday could only be understood by implementing analytics and AI.

> To understand this concept mentioned above, contemplate the business email inbox, which might receive 1000s of emails every day from customers. If through AI systems or analytical processes a pattern could be identified among these emails, these emails could be easily segregated among the various departments and directed towards them, which will create value addition to the whole process and enable customer satisfaction.

Although tools such as PowerBI, Tableau, R, Python are of great importance in the field of analytics and AI, without proper quality data they are useless. In this case, quality data can be compared to the fuel of a vehicle. If the quality of the fuel put in a vehicle is not of good quality the vehicle would not be able to perform well and may also fail totally. Just in the same context the quality of data is very important to actually design robust and accurate AI/ML models to deliver high performance results. So not just data but the quality of data is also very important for any meaningful analytics.

7.3.3 Data Requires Infrastructure

Another similarity between oil and data is that both require infrastructure for storage and transport. The infrastructure required for data consists of different hardware and software with worldwide networks which act as a roadway for the transportation of data. It requires storage technologies such as DAS, NAS, and SAN concepts and hardware such as hard-drives and tape drives. There are also data servers and databases for the handling and transportation of data. Another recent technology which enables data handling today is cloud computing. So, infrastructure consisting of hardware and software are extensively used for acquiring, processing, storing, and communicating data.

The following properties are imperative in any infrastructure.

Availability: the data in the system can be retrieved within a realistic time period, as the data may be required frequently for analytics.

Fault-tolerant: the system should be able to deliver service with a high uptime and can handle failures and mitigate loss of data due to several reasons. A few methods which are employed in this area are RAID Technology, distributed computing, and built-in redundancy.

Cost-effective: with all the above properties and facilities, the system needs also to be made available at a reasonable rate.

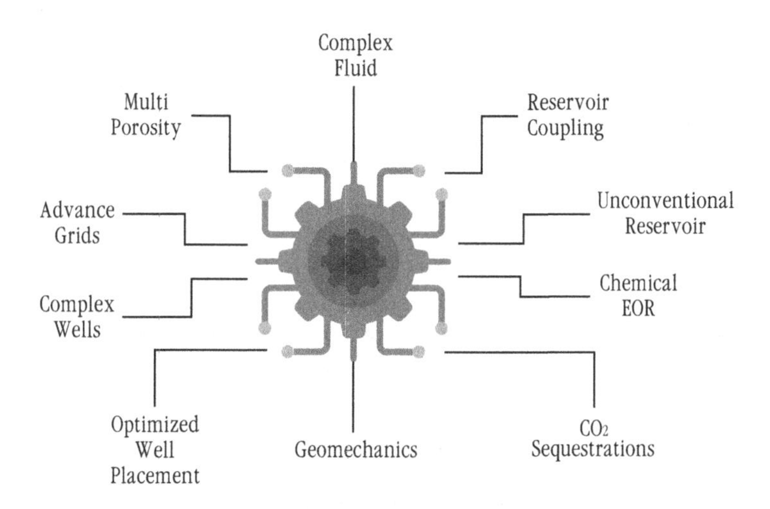

FIGURE 7.4 Simulation engine architecture.

7.3.4 Difference between Data and Oil

The most important difference between data and oil is that oil is a limited resource, but data is unlimited and when anything is abundant it becomes a big challenge to handle and use it properly to one's advantage.

Data is an asset which is ever-growing and needs to be treated with care and respect to make any predictions of market behaviour. If companies have access to 20 years of market dynamics and customer behaviour they can create highly accurate models to simulate the future market behaviour and design robust decision support systems.

7.4 CHALLENGES OF DATA INTEGRATION IN THE OIL AND GAS SECTOR

Today data is the main driving force behind any decision a company makes. And when billions of dollars are at stake it becomes paramount to take an informed decision, but there are certain challenges affecting the oil and gas sector. The most important of these are set as below:

1. High volume of data.

   The amount of data generated from the different Transaction Processing Systems are huge and ever increasing. The amount of data generated during just E&P processes is over 2 TB (Karen Boman, Rigzone Staff 2015) with 36 million sensors (Karen Boman, Rigzone Staff 2015) being used. The number of sensors being used to gather data for one process is also increasing exponentially. Thus, one can imagine just how much data would be generated in the near future.
2. Highly diversified sources and data variety.

   The data is generated from various sources such as geological and geophysical processes, seismic survey, streaming data from sensors. Again, some of this data is structured, some is unstructured or semi-structured. Handling, integrating, and managing this requires highly complex systems.
3. Incompatible data management.

   The different data which is acquired from different sources is mostly in different formats and is thus incompatible. Just the G&G data is processed through different types of software which delivers output in different formats.
4. Response time to the user needs.

   The most important part of IT systems is to deliver the results to the users at the point of time when required. If response time is high, then the usability of the results would not be of a high value to the user.

7.5 REDUCING PRODUCTION COST THROUGH IT TECHNOLOGIES

It is very important to understand how IT has been able to resolve the issues mentioned above as well as to be able to reduce production cost in the whole hydrocarbon value chain. There are four broader ways by which IT has been able to reduce production cost and increased throughput.

1. Analytics and simulations enable a company to predict and control the dynamics of the market and accordingly maintain production schedule so that breakeven on $40/barrel can be achieved and profitability can be maintained at every point.
2. A lot of investment is being made for R&D for innovative techniques to extract more crude from the existing wells as well as to find newer reservoirs. Technologies are being developed so that already-completed wells could be re-opened for production. Technologies such as GIS, SCADA, telemetry, image processing, and prediction modelling have enabled a lot of the above-mentioned criteria.

   To explain with an example, Enhanced Oil Recovery (EOR) is done by understanding various geological and geophysical parameters and then injecting high-pressure steam, chemical foams, or biochemicals. All these require a lot of data processing, analytics, and decision support systems to be able to take the proper decision, so that it does not damage the reservoir and also maximizes production. The current AI-based software can give very accurate estimates of the pressure, temperature, and any other physical parameters to increase the efficiency of the production unit.
3. The next most important technology being used is automation in different processes. Technologies such as SCADA and telemetry enable higher precision, fault tolerance, and higher rate of throughput. It also removes losses due to human error.
4. One of the biggest challenges in the hydrocarbon value chain is supply chain management, and IT technologies have been able to resolve the issues with the integration of technologies such as GIS, GPS tracking, and predictive analytics on the same platform.

7.6 CASE STUDY

Cost Reduction in the Oil and Gas Industry

The ABB's challenge is to increase the productivity at lower cost and maintain higher safety standards for the operators and to reconfigure the processes to be able to give a competitive edge over the competitor in the market. The benefits were evident within 100 days from the start of the implementation of newer technologies. The modernization process took into consideration low-value events and understanding the factors affecting the market, thereby reducing the cost incurred to lift the hydrocarbon. One of the most important and effective methods to reduce cost of production is to identify and eliminate low-value activities in engineering.

If the process is implemented the cost savings are achieved by the following parameters.

1. Maintainable safety and integrity.

2. Enhanced production efficiency/uptime.
3. Structural focused key activities.

The benefits of these changes have been maintained for years after the implementation of the processes.

The low-value activities which were identified for discontinuation are as follows:

1. Redundant equipment maintenance.
2. Frequent inspection and maintenance activities.
3. Carrying out maintenance or inspection during TARs (Time at Risk) that could be done online.

The above-mentioned parameters are only a few among others which were identified and discontinued.

- ABB's approach.

The company created the "100 day challenge" which is an approach to enhance the efficiency in the UKCS.

Step 1: Rapid assessment to identify the opportunities that exist and can be delivered in the timeframe.
Step 2: Planning and development stage where ABB revise maintenance, inspection, and testing plans.
Step 3: Implementation and putting the revised plans into practice.

ABB has facilitated in identification and redesigning the low-value activity schedules which led to reduction in cost and improved uptime. This has been designated as the "100 Days Challenge". The company has created a challenge for any operator in the O&G field to take the "100 Day Challenge" and hunt for the redundant activities and optimize their processes.

Some distinctive areas which can provide rapid benefits include:

1. **Managing redundant equipment:** identification and decommissioning of disused systems to ensure safety, reduce downtime and maintenance.
2. **Non-invasive inspection:** identification and grouping of systems which can be inspected online thus reducing the time and cost in TARs. The frequency of inspections can also be reduced.
3. **EX hazardous area inspections:** identify the procedure to reduce the regularity and scope of inspections on electrical equipment.
4. **PSV review:** enhancing the test intervals for pressure relief.
5. **Reducing the frequency and scope of TARS:** extend the interval between TARS by removing work scope from the events and changing test and inspection intervals.

6. **Safety instrumented systems testing workload reduction:** identify the process by which online testing could be increased and frequency of proof tests, without lowering the level of protection and safety in the equipment.
7. **Improving compressor reliability:** identifying the process to reduce issues in compressors frequencies and thus cutting down on breakdown maintenance and providing higher uptime.
8. **Bringing planning, preparation and scoping work on-shore:** utilization of temporal data to estimate the frequency and need for maintenance, thus creating the optimized schedule for maintenance.
9. **Remote monitoring and diagnostics:** utilizations of tools and services to diagnose problems before they cause production losses and costly breakdown maintenance.

CONCLUSION

The challenge in this domain is numerous and highly volatile, so the dependence on technology has become imperative. With digital oilfield and automation technology a lot of problems are being mitigated and/or minimized. But in the future AI is going to be the game-changer in every aspect of this field and, when coupled with quantum computing, the whole approach to resolving problems could become a different ball game.

BIBLIOGRAPHY

Aceto, G., Persico, V., & Pescapé, A. (2020). Industry 4.0 and health: Internet of things, big data, and cloud computing for healthcare 4.0. *Journal of Industrial Information Integration*, *18*, 100129. https://doi.org/10.1016/j.jii.2020.100129

Allan, G., Hanley, N., McGregor, P., Swales, K., & Turner, K. (2007). The impact of increased efficiency in the industrial use of energy: A computable general equilibrium analysis for the United Kingdom. *Energy Economics*, *29*(4), 779–798. https://doi.org/10.1016/j.eneco.2006.12.006

Anderson, S. T., & Newell, R. G. (2004). Information programs for technology adoption: The case of energy-efficiency audits. *Resource and Energy Economics*, *26*(1), 27–50. https://doi.org/10.1016/j.reseneeco.2003.07.001

Ashurst, C., Freer, A., Ekdahl, J., & Gibbons, C. (2012). Exploring IT-enabled innovation: A new paradigm? *International Journal of Information Management*, *32*(4), 326–336. https://doi.org/10.1016/j.ijinfomgt.2012.05.006

Bányai, T., Tamás, P., Illés, B., Stankevičiūtė, Ž., & Bányai, Á. (2019). Optimization of municipal waste collection routing: Impact of Industry 4.0 technologies on environmental awareness and sustainability. *International Journal of Environmental Research and Public Health*, *16*(4), 634. https://doi.org/10.3390/ijerph16040634

Bartodziej, C. J. (2017). The concept Industry 4.0. In C. J. Bartodziej, *The Concept Industry 4.0* (pp. 27–50). Springer Fachmedien Wiesbaden. https://doi.org/10.1007/978-3-658-16502-4_3

Bider, I., & Jalali, A. (2016). Agile business process development: Why, how and when—Applying Nonaka's theory of knowledge transformation to business process development. *Information Systems and E-Business Management*, *14*(4), 693–731. https://doi.org/10.1007/s10257-014-0256-1

Bolton, R. N., McColl-Kennedy, J. R., Cheung, L., Gallan, A., Orsingher, C., Witell, L., & Zaki, M. (2018). Customer experience challenges: Bringing together digital, physical and social realms. *Journal of Service Management*, *29*(5), 776–808. https://doi.org/10.1108/JOSM-04-2018-0113

Chauhan, C., Singh, A., & Luthra, S. (2021). Barriers to Industry 4.0 adoption and its performance implications: An empirical investigation of emerging economy. *Journal of Cleaner Production*, *285*, 124809. https://doi.org/10.1016/j.jclepro.2020.124809

Chen, G., Wang, P., Feng, B., Li, Y., & Liu, D. (2020). The framework design of smart factory in discrete manufacturing industry based on cyber-physical system. *International Journal of Computer Integrated Manufacturing*, *33*(1), 79–101. https://doi.org/10.1080/0951192X.2019.1699254

Chowdhury, K., Arif, A., Nur, M. N., & Sharif, O. (2020). A cloud-based computational framework to perform oil-field development & operation using a single digital twin platform. *Day 2* Tue, May 05, 2020, D021S018R006. https://doi.org/10.4043/30735-MS

Chronis, A. et al. (2022). Striking the balance: How and where will oil and gas producers deploy their cash. *Deloitte Insights*, August 25.

Costa, F., Hagan, J. E., Calcagno, J., Kane, M., Torgerson, P., Martinez-Silveira, M. S., Stein, C., Abela-Ridder, B., & Ko, A. I. (2015). Global morbidity and mortality of leptospirosis: A systematic review. *PLOS Neglected Tropical Diseases*, *9*(9), e0003898. https://doi.org/10.1371/journal.pntd.0003898

Davenport, T. H. (2014). How strategists use "big data" to support internal business decisions, discovery and production. *Strategy & Leadership*, *42*(4), 45–50. https://doi.org/10.1108/SL-05-2014-0034

DeSteno, D., Bartlett, M. Y., Baumann, J., Williams, L. A., & Dickens, L. (2010). Gratitude as moral sentiment: Emotion-guided cooperation in economic exchange. *Emotion*, *10*(2), 289–293. https://doi.org/10.1037/a0017883

Djahel, S., Doolan, R., Muntean, G.-M., & Murphy, J. (2015). A communications-oriented perspective on traffic management systems for smart cities: Challenges and innovative approaches. *IEEE Communications Surveys & Tutorials*, *17*(1), 125–151. https://doi.org/10.1109/COMST.2014.2339817

Feineman, D. R. (2014). Assessing the maturity of digital oilfield developments. *All Days*, SPE-167832-MS. https://doi.org/10.2118/167832-MS

Ghandi, A., & Lin, C.-Y. C. (2014). Oil and gas service contracts around the world: A review. *Energy Strategy Reviews*, *3*, 63–71. https://doi.org/10.1016/j.esr.2014.03.001

Grangel-Gonzalez, I., Halilaj, L., Coskun, G., Auer, S., Collarana, D., & Hoffmeister, M. (2016). Towards a semantic administrative shell for Industry 4.0 components. In *2016 IEEE Tenth International Conference on Semantic Computing (ICSC)* (pp. 230–237). https://doi.org/10.1109/ICSC.2016.58

Hughes, D. M. (2013). Climate change and the victim slot: From oil to innocence: Climate change and the victim slot. *American Anthropologist*, *115*(4), 570–581. https://doi.org/10.1111/aman.12044

Humphreys, D. (2020). Mining productivity and the fourth industrial revolution. *Mineral Economics*, *33*(1–2), 115–125. https://doi.org/10.1007/s13563-019-00172-9

Jakob, D. (2013). Crafting your way out of the recession? New craft entrepreneurs and the global economic downturn. *Cambridge Journal of Regions, Economy and Society*, *6*(1), 127–140. https://doi.org/10.1093/cjres/rss022

Kagermann, H. (2015). Change through digitization—Value creation in the age of Industry 4.0. In H. Albach, H. Meffert, A. Pinkwart, & R. Reichwald (Eds.), *Management of Permanent Change* (pp. 23–45). Springer Fachmedien Wiesbaden. https://doi.org/10.1007/978-3-658-05014-6_2

Kamble, S. S., Gunasekaran, A., & Sharma, R. (2018). Analysis of the driving and dependence power of barriers to adopt Industry 4.0 in Indian manufacturing industry. *Computers in Industry, 101*, 107–119. https://doi.org/10.1016/j.compind.2018.06.004

Kochovski, P., & Stankovski, V. (2018). Dependability of container-based data-centric systems. In *Security and Resilience in Intelligent Data-Centric Systems and Communication Networks* (pp. 7–27). Elsevier. https://doi.org/10.1016/B978-0-12-811373-8.00001-X

Korovin, I. S., & Tkachenko, M. G. (2016). Intelligent oilfield model. *Procedia Computer Science, 101*, 300–303. https://doi.org/10.1016/j.procs.2016.11.035

Larson, D., & Chang, V. (2016). A review and future direction of agile, business intelligence, analytics and data science. *International Journal of Information Management, 36*(5), 700–710. https://doi.org/10.1016/j.ijinfomgt.2016.04.013

Lasi, H., Fettke, P., Kemper, H.-G., Feld, T., & Hoffmann, M. (2014). Industry 4.0. *Business & Information Systems Engineering, 6*(4), 239–242. https://doi.org/10.1007/s12599-014-0334-4

Lennvall, T., Gidlund, M., & Akerberg, J. (2017). Challenges when bringing IoT into industrial automation. *2017 IEEE Africon* (pp. 905–910). https://doi.org/10.1109/AFRCON.2017.8095602

Li, L. (2018). China's manufacturing locus in 2025: With a comparison of "Made-in-China 2025" and "Industry 4.0." *Technological Forecasting and Social Change, 135*, 66–74. https://doi.org/10.1016/j.techfore.2017.05.028

Lins, T., & Oliveira, R. A. R. (2020). Cyber-physical production systems retrofitting in context of Industry 4.0. *Computers & Industrial Engineering, 139*, 106193. https://doi.org/10.1016/j.cie.2019.106193

Liu, Z., Xie, K., Li, L., & Chen, Y. (2020). A paradigm of safety management in Industry 4.0. *Systems Research and Behavioral Science, 37*(4), 632–645. https://doi.org/10.1002/sres.2706

Lu, H., Guo, L., Azimi, M., & Huang, K. (2019). Oil and gas 4.0 era: A systematic review and outlook. *Computers in Industry, 111*, 68–90. https://doi.org/10.1016/j.compind.2019.06.007

Monostori, L. (2014). Cyber-physical production systems: Roots, expectations and R&D challenges. *Procedia CIRP, 17*, 9–13. https://doi.org/10.1016/j.procir.2014.03.115

Moser, P., Isaksson, O., Okwir, S., & Seifert, R. W. (2021). Manufacturing management in process industries: The impact of market conditions and capital expenditure on firm performance. *IEEE Transactions on Engineering Management, 68*(3), 810–822. https://doi.org/10.1109/TEM.2019.2914995

Murry, J. W., & Hammons, J. O. (1995). Delphi: A versatile methodology for conducting qualitative research. *The Review of Higher Education, 18*(4), 423–436. https://doi.org/10.1353/rhe.1995.0008

Nagy, J., Oláh, J., Erdei, E., Máté, D., & Popp, J. (2018). The role and impact of Industry 4.0 and the internet of things on the business strategy of the value chain—The case of Hungary. *Sustainability, 10*(10), 3491. https://doi.org/10.3390/su10103491

Oesterreich, T. D., & Teuteberg, F. (2016). Understanding the implications of digitisation and automation in the context of Industry 4.0: A triangulation approach and elements of a research agenda for the construction industry. *Computers in Industry, 83*, 121–139. https://doi.org/10.1016/j.compind.2016.09.006

Palazzo, G. (2007). Organizational integrity—Understanding the dimensions of ethical and unethical behavior in corporations. In W. C. Zimmerli, M. Holzinger, & K. Richter (Eds.), *Corporate Ethics and Corporate Governance* (pp. 113–128). Springer, Berlin Heidelberg. https://doi.org/10.1007/978-3-540-70818-6_9

Pereira, A. C., & Romero, F. (2017). A review of the meanings and the implications of the Industry 4.0 concept. *Procedia Manufacturing, 13*, 1206–1214. https://doi.org/10.1016/j.promfg.2017.09.032

Perrons, R. K., & Hems, A. (2013). Cloud computing in the upstream oil & gas industry: A proposed way forward. *Energy Policy*, *56*, 732–737. https://doi.org/10.1016/j.enpol.2013.01.016

Pflaum, A. A., & Golzer, P. (2018). The IoT and digital transformation: Toward the data-driven enterprise. *IEEE Pervasive Computing*, *17*(1), 87–91. https://doi.org/10.1109/MPRV.2018.011591066

Pham, Q. T., Mai, T. K., Misra, S., Crawford, B., & Soto, R. (2016). Critical success factors for implementing business intelligence system: Empirical study in Vietnam. In O. Gervasi, B. Murgante, S. Misra, A. M. A. C. Rocha, C. M. Torre, D. Taniar, B. O. Apduhan, E. Stankova, & S. Wang (Eds.), *Computational Science and Its Applications – ICCSA 2016* (Vol. 9790, pp. 567–584). Springer International Publishing. https://doi.org/10.1007/978-3-319-42092-9_43

Rystad Energy. Upstream UCube database. Accessed September 2021.

S&P Capital IQ. Industry financials. Accessed September 2021.

8 Oil and Gas Industry in Context of Industry 4.0

An Understanding

Achala Shakya and Gaurav Tripathi

8.1 INTRODUCTION

The term "Industry 4.0" has grown in use to explain the propensity for production settings to become more automated and digital (Lasi et al., 2014). The purpose of this industry is to take a planned initiative and build intelligent factories that employ a variety of cutting-edge innovations to create human–equipment interfaces and cyber-physical systems (CPS) that result in manufacturing systems that are economically, environmentally, and socially sustainable (Oesterreich & Teuteberg, 2016; Bartodziej, 2017). These comprise additional manufacturing, big data analytics, robotics, virtual reality, the Internet of Things (IoT), cloud computing, and big data. Following Germany's initial "Industry 4.0" plan proposal, numerous nations, including Japan, the United States, and China, are now actively implementing strategic plans (Lasi et al., 2014). Germany has worked on initiatives such as "Information and Communication Technologies 2020 Research for Innovations" and "Cyber Physical Production Systems" (Lu et al., 2019). At the German Hannover Messe in April 2013, Industry 4.0 was formally introduced. This followed the Industry 1.0, 2.0, and 3.0 eras of the steam engine, electrification, and information, respectively. The objective of Industry 4.0 is to accelerate industrialization in the intelligence era by using information technology (Zhang et al., 2021). Here, Figure 8.1 shows the generation of the industries, i.e., 1.0, 2.0, 3.0, and 4.0 and how the industries have evolved. The invention of the steam engine led to the first industrial revolution, which was followed by the second's focus on mass production, the third's on computers, and the fourth's, which is distinguished by improved networking and the computerization of all production-related tasks. Nowadays, both the physical and virtual worlds make extensive use of the IoT, that encouraged organizations and governments to embark on an evolutionary road towards Industry 4.0 (Nagy et al., 2018). Artificial intelligence is integrated with production, image, and data equipment in order to achieve multidimensional intelligent production (Wang et al., 2018).

The advent of digital manufacturing, or the "smart factory", which demands for the adoption of novel business models, cognitive interconnection among enterprise divisions, fluidity in operations, versatility of industry applications, and their

DOI: 10.1201/9781003357872-8

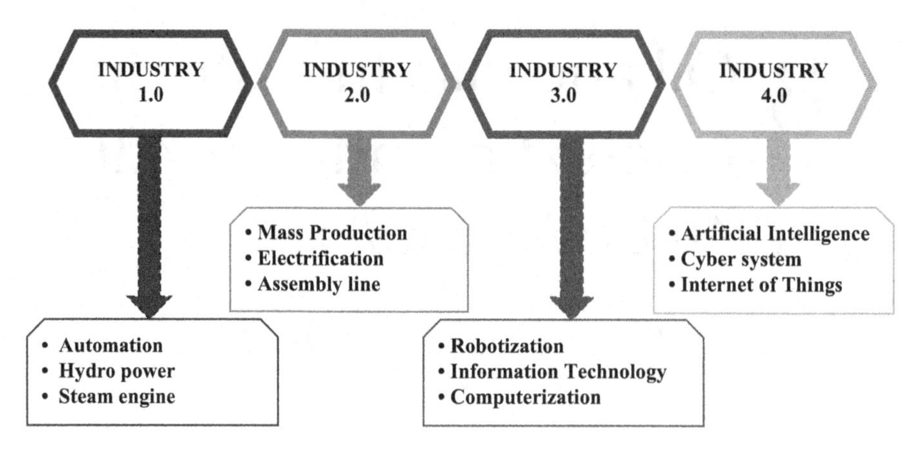

FIGURE 8.1 Generation of industries.

compatibility, forms the basis of Industry 4.0. Smart logistics makes use of intelligent technology that is integrated to give the logistics system the ability to judge and resolve logistics-related issues through perception, learning, and reasoning (Wang et al., 2018). Additionally, intelligent integration technologies are used by smart services to automatically identify users' explicit and implicit needs and proactively and effectively address those needs.

Industry 4.0 has nine key technologies that have been identified as its pillars (Vaidya et al., 2018). system integration, big data, augmented reality, digital twins, self-driving automobiles, and the Internet of Things. This contributes to an innovative, ingenious improvement in coordination and cooperation between all of the production units, logistics, revolutionizing the manufacturing process and enhancing services delivery, planning of resources in a fruitful and economical manner (Bartodziej, 2017).

Industry 4.0 may therefore offer intelligent service and ensure that procedures are faultless in all sectors, including agriculture, healthcare, governmental organizations, manufacturing, higher education institutions, logistics, etc. Using technology like big data, 3D printing, mobile devices, and cloud computing, businesses can also empower their customers by addressing their requirements, creating a new intelligent environment. Industry 4.0 is also referred to as "Industry 4.0" in German and "lanouvelle france industrielle" in French and "Advanced Manufacturing Partnership" in the United States (Grangel-Gonzalez et al., 2016; Vaidya et al., 2018). Industry 4.0 is a first step towards an intelligent world and has a wide range of applications. A systematic review of academic articles was conducted by the authors to study different elements of Industry 4.0. This review was based on all the theories and data presented above. The strategy framework for Industry 4.0 is depicted in Figure 8.2 (Lasi et al., 2014; Lins & Oliveira, 2020), which comprises four primary themes: smart factories, manufacturing, logistics, and services.

Products produced by Industry 4.0 result from the close integration of industrialization and digitalization (Kagermann, 2015). One of them is the "digital factory",

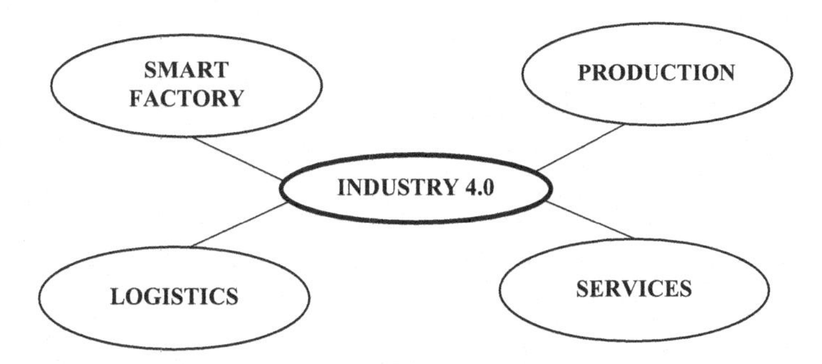

FIGURE 8.2 Major themes of the industry 4.0.

which combines big data, monitoring, and IoT methods to improve management services and make the manufacturing process more controllable. The idea of an intelligent factory is based on the digital factory; as a result, image and equipment data are combined with artificial intelligence to achieve multidimensional intelligent manufacturing (Chen et al., 2020). Utilizing intelligent technology that works together, smart logistics gives the logistics system the capacity to assess and decide on specific logistics-related challenges through perception, learning, and reasoning. Additionally, intelligent integration technologies used by smart services automatically identify the users' demands explicitly and implicitly and effectively address those needs. Following Germany's initial "Industry 4.0" plan proposal, numerous nations, including China, Japan, and the United States are now actively implementing development goals (Lasi et al., 2014). Germany has worked on initiatives including "Information and Communication Technologies 2020—Research for Innovations" and "Cyber Physical Production Systems" (Chen et al., 2020; Monostori, 2014). China developed the "Made in China 2025" initiative influenced by Germany's "Industry 4.0" and the two countries worked closely together (Li, 2018; Lu et al., 2019). China's State Council formally released "Made in China 2025" in May 2015 and put forth a detailed plan to support the development of industrial power (Xinhua et al., 2017). The re-industrialization of the US, including software, big data, and the Internet, is receiving more attention. In order to create a highly open, international industrial network, the United States has broken down the walls that separate conventional networks and businesses. It has also used big data analytics to reshape manufacturing. In addition, Japan's "White Paper on Manufacturing Industries", published in 2015, is a crucial step in its quest to dominate the global industrial value chain (Pereira & Romero, 2017). Table 8.1 shows the application areas which were triggered by the industrial revolution, i.e., Industry 4.0.

8.2 CONCEPTS OF THE OIL AND GAS INDUSTRY 4.0

The oil and gas sector is essential to the industry's management. However, due to the decline in oil prices, which has decreased by up to 70% from 2014 to the present, the oil and gas business has not been steady in recent years (Bartodziej, 2017; Lu et

TABLE 8.1
Application Areas under the Industrial Revolution 4.0

Area	Application	Reference
Exploration	Geological information, seismic interpretation, and geological maps in 1D, 2D, and 3D.	(Humphreys, 2020)
Drilling	Productivity improvement, NPT (non-production time) reduction, and risk mitigation. Defining the dynamics of the drill string.	(Tran Vu Tung et al., 2020)
Reservoir Engineering	Closed-loop reservoir management (CLRM), reservoir management, Integrated Asset Modelling (IAM), optimizing heavy oil reservoirs, unique reservoir characterization, enhanced oil recovery projects and improved hydraulic fracturing.	(Lu et al., 2019)
Production Engineering	Enhanced assessment of the decline curve, production back allocation. (Rod pump optimization, more effective hydraulic fracturing, enhanced reservoir monitoring.)	(Tran Vu Tung et al., 2020)
Maintenance	Optimizing asset management and well completion.	(Lu et al., 2019)
Health and Safety Environment—HSE)	Effective HSE management, improved risk assessment method.	(Liu et al., 2020)

al., 2019). Additionally, the output of oil and gas decreased by 3% to 5%, several oil and gas companies discovered that it was already challenging for them to keep their trade balance, and an increasing number of businesses are either being purchased or would soon confront the difficulty of being sold. Mass layoffs were implemented by some businesses to cut costs, resulting in a 40% increase in the number of unemployed persons in some industries and a substantial decline in employment rates across the board, including the oil and gas sector (Allan et al., 2007). These facts show how fundamentally the drop in oil prices is influencing the sector. Moreover, given that supply and demand relationships play a significant role in determining oil and gas prices and that these relationships exhibit cyclical behavior, the state of the oil and gas industry at present is one of challenging transition. Because of supply and demand fluctuations and a lower total return to shareholders (TRS) than other industries, investors are cautious of the oil and gas sector. The fluctuating trends in supply and demand are also having an impact on the oil and gas industry. Several factors, including production rates, well testing, fluid analysis, seismic surveys, etc., are shaking up the oil and gas value chain.

The oil and gas sector has decreased its effectiveness over the years, outlining technical constraints, in contrast to the majority of other asset-intensive businesses (Moser et al., 2021). In this context, there is a lot of pressure on oil and gas companies to accelerate their quest for optimization in order to lower costs, boost output, and boost profits. As a result of Industry 4.0's rapid developments, old industrial

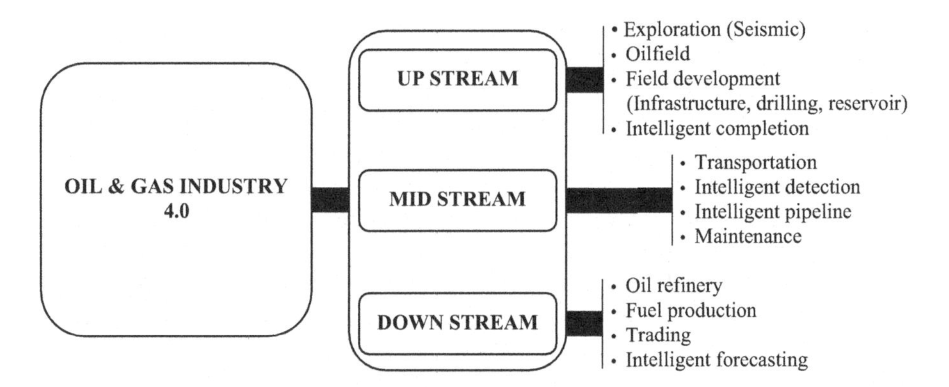

FIGURE 8.3 Taxonomy of oil and gas industry 4.0.

production techniques will gradually change and be replaced by new ones (Aceto et al., 2020). Examples of these new technologies include cloud computing, big data, and the IoT. Therefore, the oil and gas business must rely on Industry 4.0 to flourish.. Earth-shattering developments in the growth of the global oil and gas industry will occur every decade. Utilizing cutting-edge digital technologies to increase value in the industry is "oil and gas 4.0's" key goal. However, several companies are taking a long time to digitalize (Lu et al., 2019). As per data, one-third of oil and gas corporations claim that their use of digital technology is "new" or "exploratory". Normally, the oil and gas sector should be at the forefront of new technology adoption, but in practice, only a small percentage of the sector can do so (Anderson & Newell, 2004). The oil and gas business has used a variety of technology recently, including robots and satellites. However, it can be claimed that there is no interdisciplinary integration because all these technologies are asset-level (Lu et al., 2019). The oil and gas business has a digitalization score of 4.68 (0–10) in Deloitte's 2015 assessment (Costa et al., 2015). Only a limited number of top businesses have advanced to a top standard of digitization and are advancing towards positive changes (Lu et al., 2019). The petroleum (or oil and gas) industry is composed of three segments: upstream, midstream, and downstream. The terms "upstream" and "midstream" refer to the exploration, development, and conveyance over distant locations of oil and gas, respectively; "downstream" refers to the refining, production of fuels, polymers, and lubricants, as well as the sale of oil and gas products. Additionally, we have emphasized the applications of AI and their potential outcomes in the segments of the oil and gas industry 4.0 (Figure 8.3).

8.3 CONCEPTS OF DIGITAL OILFIELDS

"Intelligent oilfields" essentially allow for the simultaneous growth of oil and gas field advancement, decision-making, production, processing, and contemporary applications in information technology. The intelligent oilfield is more sophisticated than the digital oilfield from the perspective of applications because it substitutes the manual labour of humans as well as analyses of their work., this is also known as the knowledge-creation process. The knowledge-creation process includes awareness about the problem statement, analysis, decision-making, execution, and finally optimizing the decision to reach the goal state (Korovin & Tkachenko, 2016). Ghandi and Lin proposed the general framework of the intelligent oilfield in the year 2017 and to be able to implement intelligent oilfield, several layers of software and hardware needs to be integrated together such as resource, service, transmission, platform, application, and perceptual layers (Ghandi & Lin, 2014). Digital oilfield focuses on the incorporating of digital innovations into the workflows and processes of the oil and gas sector. Recently, they have expanded vigorously, and various multi-national/national companies are operating on it. As an illustration, major British oil companies such as British Petroleum (BP) employ automation and sensors to transfer real-time data collected on the ground to a remote centre for processing, enabling faster on the basis of analysis and investigation (DeSteno et al., 2010).

In order to facilitate multidisciplinary, remote cooperation across several sites, they have also constructed "Advanced Collaboration Centers" all around the world. At the commencement of the project's deployment, a business component with three levels that included digital architecture, information technology, architecture management of distant performance, and system optimization was designed. Furthermore, Shell's "Smart Fields" technology connects sensors and control valves in difficult reservoir conditions to best use of the available through ongoing observation (van den Berg et al., 2007). Major corporations predict that digital oilfields can boost output by up to 7% while lowering operating expenses to 5%. The procedure of digital oilfield developments also dictates that digital oilfields should be able to perform the following tasks: (1) sharing and accessing of real-time data, (2) capacity for evaluating the current situation, provision of forecasting correlations, and availability of optimizing the decision-making; (3) capability of achieving operational inclusion; and (4) capacity for automatic control.

8.4 CLOUD INTEGRATION IN THE OIL AND GAS DOMAIN

Oil and gas domain encounters many issues related to data integration, accomplishment sharing, and team cooperation (Bartodziej, 2017). IT is envisioned as a system that ushers in the future era by integrating big data for improved decision-making, optimized systems for superior economics, and cloud to redesign company operations and encourage innovation. With designing an IT infrastructure that is data-driven, task-focused, and cloud-managed, any business can enter this new era. Similar to utility computing, cloud computing makes use of pooled resources, information, software, and other necessary devices (Kagermann, 2015). It operates on a paid basis as per the model usage and provides hosted services online. It focuses on expanding current capabilities of IT without any hardware or software set up requirements on the business location. The cloud is altering how IT purchases its resources and vendors create their goods. These days business organizations are integrating cloud in oil and gas because of several reasons, like reduction of cost, adaptability , optimum utilization of resources, and fast response to business requirements (Feineman, 2014). Cloud is also known as virtual computing as it enables virtual processing and is capable of handling Big Data at a very low cost. Cloud computing takes all the advantages of distributed computing as well as clusters. With the help of cloud computing, real-time cooperation for reservoir models for various projects and data access from scalable systems is possible. Even though the cloud is widely used in various applications, the oil and gas industry has not still fully adopted cloud computing. Presently, out of 100%, only 10.3% of oil and gas industries are adopting or deploying cloud computing, and the remaining 7% are using cloud computing for developing the technological road map, according to the survey of IDC's 2010 Vertical Group (Lennvall et al., 2017).

It is essential for the individual user to communicate swiftly and effectively with data collecting, processing, and interpretation anywhere, at any time if the organization is to succeed in the competitive energy market of today. That is not a simple task. An industry that is now experiencing some of the most severe management issues on the planet would benefit greatly from this new strategy. Exploration and production are now more difficult than ever because of the wild fluctuations in supply and demand, unstable prices, and changing global energy policy. Energy corporations are also grappling with the global recession, just like businesses in almost every sector (Jakob, 2013). Despite all odds, cloud computing is the next-generation tool for the oil and gas industry.

The end users must be able to interact effectively everywhere with data collection, processing, and analysis to succeed the organization in the competitive market today. This new strategy is timely for a sector that is experiencing some of the most severe management issues on Earth (Grangel-Gonzalez et al., 2016; Lins & Oliveira, 2020). Exploration and production are more complicated now than they have ever been due to wild fluctuations in demand and supply, unstable prices, and changing global energy regulations. The global recession is a problem for energy corporations as it is for businesses in almost every sector. Without a doubt, the next-generation instrument for the oil and gas industry is cloud computing (Pereira & Romero, 2017; Perrons & Hems, 2013).

Organizational change includes accessibility, privacy, fast graphic visualization, and energy companies must have immediate access to several software applications interlinked by cloud-based platforms, across numerous displays. A relational database hosted on the web, a cloud services operating system, and strong connection and interoperability options are all included in Microsoft's new Windows Azure platform. These features facilitate the transition to a hybrid setup of on-premises applications and web services. This strategy includes a collection of web, server, and client-side development tools (Yang & Liu, 2012). It offers pay-as-you-go pricing, support, and service-level agreements' security along with the ease of automated service management. Additionally, it builds a cloud that can adapt to the rapidly changing business requirements of the energy industry. This cloud can scale up or down effortlessly and automatically. Today, businesses in the oil and gas industry, including energy companies and technology suppliers, are utilizing collaboration. Recent developments in exploration computing provided on-premises or online were made possible by iStore (the Information Store), which recently unveiled the first digital oilfield visualization solution in the industry via a public cloud (Chowdhury et al., 2020).

8.5 DIGITAL TRANSFORMATION

Digital platforms are affecting enterprises as well as practically every aspect of human life. As per today's scenarios, businesses must incorporate digital technology and their capacity to change practices, encourage participants, and develop new operational business models in order to compete, operate, and flourish in this digitalized era (Bolton et al., 2018). A digital transformation is to incorporate transformation that digitization and digitalization processes have made possible. Instead of producing fundamentally new software technologies, a digital revolution leads to the making of innovative software technology with their applications (Perrons & Hems, 2013). Since the core of the digitization is the modification of organizational operations that are made possible or motivated by the digitalization of technologies, the term "digital" can be misleading (Pflaum & Golzer, 2018). Digital transformation aims to work in a different manner to reach the goal by developing innovative operational and business paradigms reinforced by digital technologies to promote productivity as well as the innovations within an organization (Ashurst et al., 2012). It goes beyond automating processes or adding technologies into them to maximize the present value stream. Corporations would be able to offer new services and improved manners of providing clients with standards and achieving a competitive edge by changing the way business is performed (Ashurst et al., 2012). In the current global business environment, the capability to respond quickly and successfully to developments is among the most significant factors for success. The most current recession in the oil and gas industry serves as an illustration of this. Oil and gas firms might use technology during the upturns in economic cycles through digital business transformation and management, which will also boost their operations during the economic crises. Oil and gas businesses may be able to lessen the effects of a downturn on their business by embracing rig mechanization. As region-specific details may be recorded, rig mechanization would enable dropping and selecting new rigs through areas. This results in operations that are more durable, reliable efficiencies, and a future decrease in the requirement to re-learn the exploration difficulties. It would also improve the bottom line and improve revenue. The world economy is transforming to a digital world and economy as a result of the development of cloud applications, data analytics, smartphone, Internet connections, e-business, social networks, using intelligent sensors and IoT (Priyanka et al., 2021). Offering a seamless customer service throughout all organizational crossings ought to be the goal (Bolton et al., 2018).

Digital transformation has been a source of continued entrepreneurship and corporate dynamism, particularly in the technologically advanced oil and gas industry. Digital change has fuelled ongoing initiative and business mobility, notably in the technically sophisticated oil and gas industries (Wang & Ahmed, 2007). For businesses to successfully adapt, they must reconfigure themselves in order to work in two distinct ways, conventional and destabilizing. While the destabilizing mode looks for "additional opportunities to exploit new markets and innovate in technologies, processes, products, or services", the standard mode continues to run conventional businesses and operations. In order for businesses to succeed in the digital era, business must go through two major digital initiatives: digitization and digitalization (Vaidya et al., 2018; Wang & Ahmed, 2007). In order to properly discuss digital transformation, it is necessary to first define and consider the distinctions between these two concepts. In literature, these words are frequently used interchangeably, and many people fail to recognize the significance of each. Regarding digital transformation, the phrases have significantly distinct meanings and ambiguities, so this interchange has perplexed academics. Although adopting new technologies such as analytics, artificial intelligence, and the cloud are requirements for both digitization and automation, becoming automated and digital are two distinct initiatives that call for the application of very different rules and strategies according to the MIT Center for Information Systems Research's most current research briefing (Saarikko et al., 2020). Digitization is one of the main objectives and the fundamental component of the oil and gas Industry. It consists of the conversion of a variety of intricate information into binary code that a machine can understand. Digitization has the potential to dramatically enhance efficiency, minimize time, and improve security measures. Oil and gas firms began using digital technology in the 1980s to increase the global average functioning of oilfields, to better comprehend stocks, and overall output of oil and gas resources. However, for most of the last decade, the industry has still not comprehended the chance to leverage technology and information.

For instance, a single oil rig can produce terabytes of data each day, yet only a small part of that data is used to make decisions. Thus, digitization remains in its beginning phases for the oil and gas business. Insiders in the sector are prepared for digital innovations that the industry has yet to adopt, which is one of the causes. Digital technology has been well known over the previous ten years. The four themes that make up the oil and gas sector's digital transformation are "circular collaborative ecosystem", virtual asset life cycle administration, "beyond the barrel", and "energizing new energies". Each theme has related activities and relevant technologies. By gathering data, virtual asset life cycle administration improves strategic decisions and changes operational models. It initially employs specialized sensors to capture real-time information from physical resources. Then, it processes the data using cloud analysis tools and uses the data to understand how it affects the workflow's other steps. Finally, it makes inferences based on the information and applies lessons learned in the future. The circular collaborative ecosystem accelerates innovation, lowers costs, and increases operational transparency while fostering better collaboration among oil and gas industry participants. Modern client participation models can boost flexibility and adaptability, which can lead to new prospects for oil and gas companies. This is referred to as going "beyond the barrel" (Palazzo, 2007). "Energizing new energies" refers to the application of digitalization to managers' pursuit of energy-optimized innovations and the advancement of innovative generation.

8.6 AGILE BUSINESS TRANSFORMATION

Presently, Business Intelligent (BI) software and hardware are part of big data technologies (Davenport, 2014). Making information intelligently usable is the primary goal of Business Intelligence. The scope of the BI project is concentrated on converting data into information if the major objective of Business Intelligence is to facilitate the utilization of information. The process of converting data into information includes software development; however, in Business Intelligence, software development focuses more on applying business context to data than on building a functional program. Software used in Business Intelligence includes database management systems, data cleansing, data transformation, and analytical systems (Bider & Jalali, 2016).

In contrast to the coding of a program, the scope of Business Intelligence creation includes an additional setup for the application. Information Technology will need to understand the business data usage to further implement and apply logic as well as setup the software (Larson & Chang, 2016). The issue of ambiguous requirements arises from the fact that the supply of Business Intelligence solutions frequently involve a cycle of discovery and refinement on the part of the customer. Even with the help of subject matter experts, converting data into information is not an easy procedure. For Business Intelligence, certain questions shall be taken into consideration such as data sources, data usage, data creation, and data conversion into the relevant information.

Business Intelligence systems are made up of many different components, some of which include source systems, extract, transform, and load (ETL), databases, and front-end tools. Understanding the customer need and expectations are the main requirements of Business Intelligence (Wu et al., 2007). It would be difficult to use contracts in Business Intelligence without any predefined expectations. A baseline of assumptions that permits improvement and modification is necessary for Business Intelligence projects. The goal is to put more emphasis on teamwork than on spending time developing a precise strategy. The creation of detailed plans is frequently challenging since only high-level planning information is available. Collaboration facilitates a resolution by defining expectations and improving stakeholder communication. The challenges of Business Intelligence systems are addressed by the Agile concepts of "customer cooperation over contract negotiation" as well as "responding to change over implementing a strategy". The first goal of fast analytics initiatives is speedy data collecting to be utilized for discovery (Davenport, 2014). The most well known are Scrum, Agile Data Warehousing, and Agile development methods for Business Intelligence (Djahel et al., 2015). Extreme Scoping and Agile data warehousing work well when a data warehouse is involved. Data warehouses offer a central repository with integrated data for analysis and are a crucial component of a Business Intelligence system. The data integration portion of Business Intelligence projects is covered by the Agile technique known as Extreme Scoping (Djahel et al., 2015; Pflaum & Golzer, 2018; Wu et al., 2007).

The integration, collection, and transformation of data sources into the central repository are all centred on data management tasks. The primary goals of planning processes are to gather and comprehend information gathered from various sources, cleaning, and modelling the data, and obtaining the correct data for loading. Extreme Scoping encompasses all data management operations and is "data-centric" (Kochovski & Stankovski, 2018). The process of Extreme Scoping consists of seven steps:

1. Identifying the project requirements.
2. Dividing the project requirements into several versions.
3. Concentrating on the business value of the requirements.
4. Identifying digital innovations.
5. Adding those digital innovations into the project.
6. Describing the resources needed to perform the project.
7. Adding numerous data development paths for the data deliverables.

In fact, there are several related techniques that fall under the umbrella of Agile data warehousing. For instance, one method uses an end-to-end strategy to offer an Agile data warehouse, while another method uses Agile methodologies for each stage of delivering Business Intelligence, including a specific Agile methodology for data modelling. The resemblance of all the strategies includes vision of the architecture, data prototype, modelling, anticipating multiple data definitions, work organization as per customer requirement, and involvement of the stakeholders (Hughes, 2013). For the delivery of Business Intelligence projects, data warehousing in Agile incorporates several of the same principles as in Scrum. Along with being heavily used in Agile technique for Agile software development, Scrum is also commonly employed in Business Intelligence (Kochovski & Stankovski, 2018). The product backlog, daily Scrum, sprint, sprint backlog, and user story are the main Scrum concepts used in Business Intelligence. A Business Intelligence project is the collection of Agile stories that are assembled from the Business Intelligence requirement specification. Each narrative is created, refined, reviewed, and published. A sprint is a one- to two-week cycle that comprises objectives, assessment, architecture, implementation, and end user acceptance testing. Stories are divided into a sprint backlog and a product backlog. The sprint backlog is the activity that the project team accomplishes in the sprint, whereas the product backlog is a catalogue of each story arranged by precedence that will be investigated for the subsequent sprint. Customers are consulted during each stage of the sprint to determine their needs. Despite the fact that current research shows that incorporating Agile techniques into Business Intelligence programs can enhance quality, cut down on total time, and increase client satisfaction (Kochovski & Stankovski, 2018). However, there is still the failure of Business Intelligence projects up to 60–70% (Pham et al., 2016). As per today's requirements, Agile has gained more attention due to the advancement of big data and the growing emphasis on analytics, even though some of its successes have boosted Business Intelligence adoption (DeSteno et al., 2010; Feineman, 2014).

8.7 DIGITAL LEADERSHIP

Several oil and gas operators have started digital leadership programs to increase their overall organization's knowledge of electronic technological innovations and aid in the creation of more robust strategic business plans that will eventually result in a shared vision that will lead to strategic synchronization. It is to be that not every employee inside a company needs to be a data scientist as different positions within the organization play different roles to upgrade the growth of the organization with their strategy (Costa et al., 2015). Data literates are capable of thinking about and acting on data, but they lack data scientists' expertise (Kochovski & Stankovski, 2018). The establishment of a data-driven organization involves data literacy with other related factors to obtain the results faster. Data maturity is an important component of data literacy and making several decisions based on data is the ultimate objective (Kagermann, 2015). Even though the term "data literacy" has gained popularity, more needs to be performed in order to create a data culture. The data science expert claims that "genuine data literacy should empower one to think and perform intelligently begin by comprehending the actual business problem and apply cognitive findings to tackle the future problems" (Lasi et al., 2014). Data specialists claim that chief technology officers are typically in charge of literacy campaigns; however, all chief executives need to support them and demonstrate the intended results. According to data science experts, data-driven approaches should be followed with leadership and drive the staff of the organization accountable from the top.

Projects promoting data literacy must start with the support of the organization's executive. Additionally, Weill et al. found that CEO digital expertise varied significantly across industries (Weill, 2008). Media, software, and telecom companies, which account for nearly one-third of senior team members, have the highest percentage of digitally knowledgeable executives (Weill, 2008). In the fields of building, entertainment, recreation, forestry, fishing, and hunting, less than 1% of management team members possess digital literacy. A small number of businesses with more technologically literate individuals have surpassed their peers in terms of performance, even though the bulk of senior management teams lack this understanding. Weill et al.'s study revealed that enterprises with digitally aware leadership teams outperformed the rest of the companies under study in terms of revenue growth, valuation, and net margins by a certain amount of percentage (Sepasgozar et al., 2020; Vaidya et al., 2018; Weill, 2008). However, it is also crucial that businesses do not necessarily need to reach a particular percentage of threshold for digitally intelligent leadership members to start achieving performance improvements (Ashurst et al., 2012; DeSteno et al., 2010; Grangel-Gonzalez et al., 2016; Monostori, 2014).

8.8 CASE STUDY

8.8.1 List of Case Studies on the Oil and Gas Industry

There are several case studies on the oil and gas industry listed below, which showed an improved operational efficiency at reduced costs by optimizing manual operations such as MOL group, Exxon Mobil, EQT Corporation, Shell International, Valero Energy Corporation, BP, Shell, Satorp, Scotia Gas Networks, Peru LNG, Tullow Ghana Ltd., Tullow Oil, etc. We are comparing different oil and gas industries in Table 8.2 based on their locations and provided services.

TABLE 8.2
Oil and Gas Industries

S. No.	Industries	Location	Description
1.	MOL group	Europe	1. Provides intelligent services. 2. Integrates enterprise and operational risk management.
2.	Exxon Mobil	Latin America	1. Delivers capital project. 2. Minimizes the risks, including environmental, as well as the social risks of their investment.
3.	EQT Corporation	North America	1. Provides sustainability in business and climate change. 2. Produces natural gas.
4.	Shell International	Latin America	1. Offers advice on investigative work in the areas of sustainability, health, safety, and the environment (HSE). 2. Produces ethanol in abundance.
5.	Valero Energy Corporation	North America	1. Keep record of the emissions of greenhouse gases. 2. Also provides sustainability in business and climate change.
6.	BP	Asia Pacific	1. Delivers capital projects. 2. Collection of the gas from the reservoir and then processes it through LNG (Liquified Natural Gas).
7.	Shell International	Asia Pacific	1. Distribution of ethanol. 2. Sustainability development.
8.	Satorp	Middle East	1. Risk assessment. 2. Making grassroot refineries.
9.	Scotia Gas Networks	Europe	1. Management of portfolios. 2. Assessment of environmental risks.
10.	Peru LNG	Latin America	1. Preparation of ecological management of a particular region. 2. Liquefaction of Natural Gas.
11.	Tullow Ghana Ltd.	Africa	1. Assessment and management of social and environmental impact. 2. Development of gas and flagship oil.
12.	Tullow Oil	Asia Pacific	1. Involves risk and safety studies. 2. Reduces safety issues.

Source: www.erm.com/industries/oil-gas/case-studies/

8.8.2 Analyzing the Barriers to Adopt Industrial Revolution in India

The authors tried to identify important obstacles to adopt Industry 4.0 in the manufacturing sector across India and made suggestions on what steps industry practitioners and policy officials could take to overcome these challenges.

Industry 4.0 is a significant subject of study and development since it enhances industrial outcomes and infrastructure by incorporating technological developments in manufacturing sectors (Raj et al., 2020). However, organizations still have a long way to go before successfully and on-time implementing digital principles (Chauhan et al., 2021). Significant investment expenditures and ambiguous cost–benefit ratios for industrial revolution 4.0 application sectors are the primary causes of the concerns. Additionally, the labour force misses the experience necessary to handle impending automation and Industry 4.0 implementation guidelines lack clarity, which has led to confusion in many firms. Before being successfully adopted wholeheartedly by all companies, this digital shift will need to overcome numerous obstacles (Senna et al., 2022; Sepasgozar et al., 2020).

The barriers to the adaptation (BTAs) for Industry 4.0 are identified in this study in terms of the available literature, insights of industry and academic experts. The BTAs were chosen using a two-step procedure (Yang et al., 2018). Based on the literature research, 15 BTAs were found in the first phase. The review considered Industry 4.0 literature including Scopus, Web of Science, and Emerald Insights. For even more details on Industry 4.0 and the obstacles preventing its deployment, several trade magazines, business papers, and media stories were also recommended. Some of the relevant keywords utilized as follows: "Industry 4.0", "digital manufacturing", "smart manufacturing", "potential and obstacles", "benefits and drawbacks", and "implementation" (Bányai et al., 2019; Bartodziej, 2017). The second phase involved the validation and further refinement of this group of 15 barriers by a team of 14 specialists from various backgrounds, which eventually resulted in the final selection of 12 BTAs for smart manufacturing that were appropriate for the Indian manufacturing enterprise. The experts included two senior executives from the automobile industry who worked in the industrial sector, three IT senior engineers, two senior automation engineers, and two senior engineers from the field of telecommunications. The team of specialists also included senior managers from the field study and supply chain management. Previous studies have shown that 5–50 experts are an appropriate quantity for doing qualitative research (Bányai et al., 2019). The right number of experts for the qualitative decision-making process can be between 10 and 30, according to Murry and Hammons (Murry & Hammons, 1995; Senna et al., 2022). It was thus appropriate to choose 14 specialists.

Twelve challenges to implement Industry 4.0 are found in the current analysis. The adoption of Industry 4.0 in many nations may be significantly impacted by a few challenges that weren't necessarily anticipated. Table 8.3 includes a brief summary of the BTAs for Industry 4.0.

Based on the fuzzy MICMAC (Matriced' Impacts Croise's Multiplication Applique'e a' un Classement) assessment, significant barriers were classified and a hierarchy has been developed as well based on the priority is discussed below in Table 8.4 (Kamble et al., 2018).

Future research may be done in other nations to look at the existence of different barriers and assess how they interact. The examination of the barriers by the authors is subjective, and any prejudice on the part of the individual evaluating the barriers will affect the outcome. According to the study, it might be advised to conduct empirical studies based on a survey that includes data from many businesses

TABLE 8.3
Barriers in Adaptation for Industry 4.0

Sl. No.	Adoption barrier to Industry 4.0	Description
1	Workplace interruptions	It is described as the changes in employment brought about by new technology and automation. Current manufacturing employment is susceptible to automation, which will lead to a loss of human labour. The remaining manufacturing positions will be more knowledge-based work, short-term tasks that are difficult to plan, and tasks that require quick decisions.
2	High cost of implementation	This is a reference to the capital costs that businesses will have to pay when building out their Industry 4.0 infrastructure. Due to a shortage of funding for appropriate technology, it is difficult for small and moderate sized firms to adopt Industry 4.0. Emerging technologies, IoT, always pose a significant risk to businesses' financial investments because there is a chance of probable financial loss and a lack of investment recovery.
3	Process and organizational changes	This refers to the process adjustments that the introduction of CPS and smart factories will bring about. Due to automation, organizational functions may shift. Industry 4.0 has led to the creation of decentralized structures. Decisions will be taken on the shop floor. The external and internal connectivity of vertical, heterogeneity, and closed technologies is the main impediment that IoT systems and services face, even though they are advantageous across industries. Most of the firms identify these as significant concerns.
4	Demand for improved skills	The success of an extremely creative firm will depend heavily on the capabilities and competences of the personnel with the advent of Industry 4.0. Companies should put their attention into developing qualified employees under the direction of human resources management. IoT solution design and deployment efficiency necessitate extensive background knowledge from a range of technical and non-technical fields.
5	Data management techniques are lacking	Knowledge management systems are information technology solutions that improve the knowledge management process by fostering interaction, locating knowledge assets, mining knowledge repositories for untapped knowledge, storing and acquiring information, and more. IoT analysis and incorporation of inbound big data through IoT systems could benefit knowledge management; however, the current systems might not be able to handle the real-time data.
		To transition from operational to more strategic jobs, one needs to possess comprehensive technical skills. Soon, data analysts who can add value through predicting and improvement will be required.

(Continued)

TABLE 8.3 (CONTINUED)
Barriers in Adaptation for Industry 4.0

Sl. No.	Adoption barrier to Industry 4.0	Description
6	Lack of understanding of IoT	A network of physical things that are outfitted using connections, circuitry, algorithms, controllers, sensors, and other components for data transmission and reception makes up IoT. Without a doubt, the possibility of monetary benefit influences the adoption of IoT in new firms. Enterprises must, on the other hand, clearly understand the subtleties of IoT deployment, including value creation, delivery, and capture. If businesses are going to think about a business model for IoT applications, they should examine the variances accordingly. Several IoT technology and implementations are already in the initial phases of development. The statistical expenditure ratios that could be generated using IoT remain uncertain. Therefore, IoT installation is considered to be a major barrier that most industries must overcome.
7	Lack of Reference Architecture and Standards	Building and selecting an Industry 4.0 architecture has proven challenging for several applications, especially when considering wireless sensor networks. There are currently no standards or conceptual models for Industry 4.0 because of its recent concept. One important obstacle to the successful deployment of IoT, for instance, is the creation of an optimal IoT architecture within supply chains that comprises the cloud services, assets, networking, and applications layers.
8	Lack of IT equipment and Internet access	Lack of the IT infrastructure required to support the implementation of this industry. The Industry 4.0 framework's IoT is a crucial element. A major impediment to a collection of products and services could be the absence of adequate communication and signal coverage solutions. Due to inadequate signal coverage, signal attenuation occurs in many production sites.
9	Privacy and Security Concerns	On the Industry 4.0 platform, a significant volume of information is transferred, creating concerns about security and confidentiality. Employees that work virtually on servers or other platforms must be familiar with cyber security. A crucial element of the CPS is security. Because of recent IT advancements, CPS is increasingly susceptible to cyberattacks. Concerns around access to the system, applications, network, and data, together with identification and certification, protection continue to pose serious issues for organizations.
10	Problems with compatibility and seamless integration	These are the issues that could arise from updating the machines and gear that are now used in the sector. One of the key obstacles to the adoption of IoT solutions to build a cyber-physical infrastructure for IoT ecosystems is building perfect interoperability and integration among advanced techniques and information systems.
11	Issues with regulatory compliance	The goal that firms strive to attain when they make sure they are aware of and follow the necessary steps to adhere to the relevant legislation, guidelines, and standards is regulatory compliance. Organizations need stricter rules for using equipment, keeping hours, and protecting IT.

(Continued)

TABLE 8.3 (CONTINUED)
Barriers in Adaptation for Industry 4.0

Sl. No.	Adoption barrier to Industry 4.0	Description
12	Uncertainty in the law and in contracts	The legislation is challenged by digitalization as the market becomes more competitive. Laws governing data protection, artificial intelligence liability, and standardization must be considered while establishing a digital strategy. Since the virtual organization lacks a legal personality and hence has no existence in law, it cannot be recognized as a legally autonomous body. Every virtual business that makes use of ICT must ensure that data transfers are safe, that they don't violate privacy laws, and that the contracts they've made are legitimate and enforceable.

TABLE 8.4
Hierarchy Level in Barriers for the Industry 4.0

Priority level	BTAs for Industry 4.0
High	Uncertainty in the law and in contracts
Moderate	Workplace disruptions
	Process and organizational changes
	Demand for improved skills
	Lack of information management systems
	Reference architecture and standards are lacking
	Lack of IT infrastructure and Internet access
	Protection related issues
	Problems with compatibility and seamless integration
	Issues with regulatory compliance
Low	High cost of implementation
	Lack of understanding of IoT

to look into the linkages and dependencies among the barriers. Structural Equation Modelling (SEM) may be used in future research to statistically validate the produced model.

REFERENCES

Aceto, G., Persico, V., & Pescapé, A. (2020). Industry 4.0 and health: Internet of things, big data, and cloud computing for healthcare 4.0. *Journal of Industrial Information Integration*, *18*, 100129. https://doi.org/10.1016/j.jii.2020.100129

Allan, G., Hanley, N., McGregor, P., Swales, K., & Turner, K. (2007). The impact of increased efficiency in the industrial use of energy: A computable general equilibrium analysis for the United Kingdom. *Energy Economics*, *29*(4), 779–798. https://doi.org/10.1016/j.eneco.2006.12.006

Anderson, S. T., & Newell, R. G. (2004). Information programs for technology adoption: The case of energy-efficiency audits. *Resource and Energy Economics*, *26*(1), 27–50. https://doi.org/10.1016/j.reseneeco.2003.07.001

Ashurst, C., Freer, A., Ekdahl, J., & Gibbons, C. (2012). Exploring IT-enabled innovation: A new paradigm? *International Journal of Information Management*, *32*(4), 326–336. https://doi.org/10.1016/j.ijinfomgt.2012.05.006

Bányai, T., Tamás, P., Illés, B., Stankevičiūtė, Ž., & Bányai, Á. (2019). Optimization of municipal waste collection routing: Impact of Industry 4.0 technologies on environmental awareness and sustainability. *International Journal of Environmental Research and Public Health*, *16*(4), 634. https://doi.org/10.3390/ijerph16040634

Bartodziej, C. J. (2017). The concept Industry 4.0. In C. J. Bartodziej, *The Concept Industry 4.0* (pp. 27–50). Springer Fachmedien Wiesbaden. https://doi.org/10.1007/978-3-658-16502-4_3

Bider, I., & Jalali, A. (2016). Agile business process development: Why, how and when—Applying Nonaka's theory of knowledge transformation to business process development. *Information Systems and E-Business Management*, *14*(4), 693–731. https://doi.org/10.1007/s10257-014-0256-1

Bolton, R. N., McColl-Kennedy, J. R., Cheung, L., Gallan, A., Orsingher, C., Witell, L., & Zaki, M. (2018). Customer experience challenges: Bringing together digital, physical and social realms. *Journal of Service Management*, *29*(5), 776–808. https://doi.org/10.1108/JOSM-04-2018-0113

Chauhan, C., Singh, A., & Luthra, S. (2021). Barriers to Industry 4.0 adoption and its performance implications: An empirical investigation of emerging economy. *Journal of Cleaner Production*, *285*, 124809. https://doi.org/10.1016/j.jclepro.2020.124809

Chen, G., Wang, P., Feng, B., Li, Y., & Liu, D. (2020). The framework design of smart factory in discrete manufacturing industry based on cyber-physical system. *International Journal of Computer Integrated Manufacturing*, *33*(1), 79–101. https://doi.org/10.1080/0951192X.2019.1699254

Chowdhury, K., Arif, A., Nur, M. N., & Sharif, O. (2020). A cloud-based computational framework to perform oil-field development & operation using a single digital twin platform. *Day 2* Tue, May 05, 2020, D021S018R006. https://doi.org/10.4043/30735-MS

Costa, F., Hagan, J. E., Calcagno, J., Kane, M., Torgerson, P., Martinez-Silveira, M. S., Stein, C., Abela-Ridder, B., & Ko, A. I. (2015). Global morbidity and mortality of leptospirosis: A systematic review. *PLOS Neglected Tropical Diseases*, *9*(9), e0003898. https://doi.org/10.1371/journal.pntd.0003898

Davenport, T. H. (2014). How strategists use "big data" to support internal business decisions, discovery and production. *Strategy & Leadership*, *42*(4), 45–50. https://doi.org/10.1108/SL-05-2014-0034

DeSteno, D., Bartlett, M. Y., Baumann, J., Williams, L. A., & Dickens, L. (2010). Gratitude as moral sentiment: Emotion-guided cooperation in economic exchange. *Emotion*, *10*(2), 289–293. https://doi.org/10.1037/a0017883

Djahel, S., Doolan, R., Muntean, G.-M., & Murphy, J. (2015). A communications-oriented perspective on traffic management systems for smart cities: Challenges and innovative approaches. *IEEE Communications Surveys & Tutorials*, *17*(1), 125–151. https://doi.org/10.1109/COMST.2014.2339817

Feineman, D. R. (2014). Assessing the maturity of digital oilfield developments. *All Days*, SPE-167832-MS. https://doi.org/10.2118/167832-MS

Ghandi, A., & Lin, C.-Y. C. (2014). Oil and gas service contracts around the world: A review. *Energy Strategy Reviews*, *3*, 63–71. https://doi.org/10.1016/j.esr.2014.03.001

Grangel-Gonzalez, I., Halilaj, L., Coskun, G., Auer, S., Collarana, D., & Hoffmeister, M. (2016). Towards a semantic administrative shell for Industry 4.0 components. In *2016*

IEEE Tenth International Conference on Semantic Computing (ICSC) (pp. 230–237). https://doi.org/10.1109/ICSC.2016.58

Hughes, D. M. (2013). Climate change and the victim slot: From oil to innocence: Climate change and the victim slot. *American Anthropologist*, *115*(4), 570–581. https://doi.org/10.1111/aman.12044

Humphreys, D. (2020). Mining productivity and the fourth industrial revolution. *Mineral Economics*, *33*(1–2), 115–125. https://doi.org/10.1007/s13563-019-00172-9

Jakob, D. (2013). Crafting your way out of the recession? New craft entrepreneurs and the global economic downturn. *Cambridge Journal of Regions, Economy and Society*, *6*(1), 127–140. https://doi.org/10.1093/cjres/rss022

Kagermann, H. (2015). Change through digitization—Value creation in the age of Industry 4.0. In H. Albach, H. Meffert, A. Pinkwart, & R. Reichwald (Eds.), *Management of Permanent Change* (pp. 23–45). Springer Fachmedien Wiesbaden. https://doi.org/10.1007/978-3-658-05014-6_2

Kamble, S. S., Gunasekaran, A., & Sharma, R. (2018). Analysis of the driving and dependence power of barriers to adopt Industry 4.0 in Indian manufacturing industry. *Computers in Industry*, *101*, 107–119. https://doi.org/10.1016/j.compind.2018.06.004

Kochovski, P., & Stankovski, V. (2018). Dependability of container-based data-centric systems. In *Security and Resilience in Intelligent Data-Centric Systems and Communication Networks* (pp. 7–27). Elsevier. https://doi.org/10.1016/B978-0-12-811373-8.00001-X

Korovin, I. S., & Tkachenko, M. G. (2016). Intelligent oilfield model. *Procedia Computer Science*, *101*, 300–303. https://doi.org/10.1016/j.procs.2016.11.035

Larson, D., & Chang, V. (2016). A review and future direction of agile, business intelligence, analytics and data science. *International Journal of Information Management*, *36*(5), 700–710. https://doi.org/10.1016/j.ijinfomgt.2016.04.013

Lasi, H., Fettke, P., Kemper, H.-G., Feld, T., & Hoffmann, M. (2014). Industry 4.0. *Business & Information Systems Engineering*, *6*(4), 239–242. https://doi.org/10.1007/s12599-014-0334-4

Lennvall, T., Gidlund, M., & Akerberg, J. (2017). Challenges when bringing IoT into industrial automation. In *2017 IEEE Africon* (pp. 905–910). https://doi.org/10.1109/AFRCON.2017.8095602

Li, L. (2018). China's manufacturing locus in 2025: With a comparison of "Made-in-China 2025" and "Industry 4.0." *Technological Forecasting and Social Change*, *135*, 66–74. https://doi.org/10.1016/j.techfore.2017.05.028

Lins, T., & Oliveira, R. A. R. (2020). Cyber-physical production systems retrofitting in context of Industry 4.0. *Computers & Industrial Engineering*, *139*, 106193. https://doi.org/10.1016/j.cie.2019.106193

Liu, Z., Xie, K., Li, L., & Chen, Y. (2020). A paradigm of safety management in Industry 4.0. *Systems Research and Behavioral Science*, *37*(4), 632–645. https://doi.org/10.1002/sres.2706

Lu, H., Guo, L., Azimi, M., & Huang, K. (2019). Oil and gas 4.0 era: A systematic review and outlook. *Computers in Industry*, *111*, 68–90. https://doi.org/10.1016/j.compind.2019.06.007

Monostori, L. (2014). Cyber-physical production systems: Roots, expectations and R&D challenges. *Procedia CIRP*, *17*, 9–13. https://doi.org/10.1016/j.procir.2014.03.115

Moser, P., Isaksson, O., Okwir, S., & Seifert, R. W. (2021). Manufacturing management in process industries: The impact of market conditions and capital expenditure on firm performance. *IEEE Transactions on Engineering Management*, *68*(3), 810–822. https://doi.org/10.1109/TEM.2019.2914995

Murry, J. W., & Hammons, J. O. (1995). Delphi: A versatile methodology for conducting qualitative research. *The Review of Higher Education*, *18*(4), 423–436. https://doi.org/10.1353/rhe.1995.0008

Nagy, J., Oláh, J., Erdei, E., Máté, D., & Popp, J. (2018). The role and impact of Industry 4.0 and the internet of things on the business strategy of the value chain—The case of Hungary. *Sustainability, 10*(10), 3491. https://doi.org/10.3390/su10103491

Oesterreich, T. D., & Teuteberg, F. (2016). Understanding the implications of digitisation and automation in the context of Industry 4.0: A triangulation approach and elements of a research agenda for the construction industry. *Computers in Industry, 83*, 121–139. https://doi.org/10.1016/j.compind.2016.09.006

Palazzo, G. (2007). Organizational integrity—Understanding the dimensions of ethical and unethical behavior in corporations. In W. C. Zimmerli, M. Holzinger, & K. Richter (Eds.), *Corporate Ethics and Corporate Governance* (pp. 113–128). Springer, Berlin Heidelberg. https://doi.org/10.1007/978-3-540-70818-6_9

Pereira, A. C., & Romero, F. (2017). A review of the meanings and the implications of the Industry 4.0 concept. *Procedia Manufacturing, 13*, 1206–1214. https://doi.org/10.1016/j.promfg.2017.09.032

Perrons, R. K., & Hems, A. (2013). Cloud computing in the upstream oil & gas industry: A proposed way forward. *Energy Policy, 56*, 732–737. https://doi.org/10.1016/j.enpol.2013.01.016

Pflaum, A. A., & Golzer, P. (2018). The IoT and digital transformation: Toward the data-driven enterprise. *IEEE Pervasive Computing, 17*(1), 87–91. https://doi.org/10.1109/MPRV.2018.011591066

Pham, Q. T., Mai, T. K., Misra, S., Crawford, B., & Soto, R. (2016). Critical success factors for implementing business intelligence system: Empirical study in Vietnam. In O. Gervasi, B. Murgante, S. Misra, A. M. A. C. Rocha, C. M. Torre, D. Taniar, B. O. Apduhan, E. Stankova, & S. Wang (Eds.), *Computational Science and Its Applications – ICCSA 2016* (Vol. 9790, pp. 567–584). Springer International Publishing. https://doi.org/10.1007/978-3-319-42092-9_43

Priyanka, E. B., Maheswari, C., & Thangavel, S. (2021). A smart-integrated IoT module for intelligent transportation in oil industry. *International Journal of Numerical Modelling: Electronic Networks, Devices and Fields, 34*(3). https://doi.org/10.1002/jnm.2731

Raj, A., Dwivedi, G., Sharma, A., Lopes de Sousa Jabbour, A. B., & Rajak, S. (2020). Barriers to the adoption of Industry 4.0 technologies in the manufacturing sector: An inter-country comparative perspective. *International Journal of Production Economics, 224*, 107546. https://doi.org/10.1016/j.ijpe.2019.107546

Saarikko, T., Westergren, U. H., & Blomquist, T. (2020). Digital transformation: Five recommendations for the digitally conscious firm. *Business Horizons, 63*(6), 825–839. https://doi.org/10.1016/j.bushor.2020.07.005

Senna, P. P., Ferreira, L. M. D. F., Barros, A. C., Bonnín Roca, J., & Magalhães, V. (2022). Prioritizing barriers for the adoption of Industry 4.0 technologies. *Computers & Industrial Engineering, 171*, 108428. https://doi.org/10.1016/j.cie.2022.108428

Sepasgozar, S. M. E., Shi, A., Yang, L., Shirowzhan, S., & Edwards, D. J. (2020). Additive manufacturing applications for Industry 4.0: A systematic critical review. *Buildings, 10*(12), 231. https://doi.org/10.3390/buildings10120231

Tung, T. V., Trung, T. N., Hai, N. H., & Tinh, N. T. (2020). Digital transformation in oil and gas companies—A case study of Bien Dong POC. *Petrovietnam Journal, 10*, 67–78. https://doi.org/10.47800/PVJ.2020.10-07

Vaidya, S., Ambad, P., & Bhosle, S. (2018). Industry 4.0 – A glimpse. *Procedia Manufacturing, 20*, 233–238. https://doi.org/10.1016/j.promfg.2018.02.034

van den Berg, A. E., Hartig, T., & Staats, H. (2007). Preference for nature in urbanized societies: Stress, restoration, and the pursuit of sustainability. *Journal of Social Issues, 63*(1), 79–96. https://doi.org/10.1111/j.1540-4560.2007.00497.x

Wang, C. L., & Ahmed, P. K. (2007). Dynamic capabilities: A review and research agenda. *International Journal of Management Reviews*, *9*(1), 31–51. https://doi.org/10.1111/j.1468-2370.2007.00201.x

Wang, J., Ma, Y., Zhang, L., Gao, R. X., & Wu, D. (2018). Deep learning for smart manufacturing: Methods and applications. *Journal of Manufacturing Systems*, *48*, 144–156. https://doi.org/10.1016/j.jmsy.2018.01.003

Weill, L. (2008). Leverage and corporate performance: Does institutional environment matter? *Small Business Economics*, *30*(3), 251–265. https://doi.org/10.1007/s11187-006-9045-7

Wu, L., Barash, G., & Bartolini, C. (2007). A service-oriented architecture for business intelligence. In *IEEE International Conference on Service-Oriented Computing and Applications (SOCA '07)* (pp. 279–285). https://doi.org/10.1109/SOCA.2007.6

Xinhua, L., Junquan, P., Lei, S., & Zhongbin, W. (2017). A novel approach for NURBS interpolation through the integration of acc-jerk-continuous-based control method and look-ahead algorithm. *The International Journal of Advanced Manufacturing Technology*, *88*(1–4), 961–969. https://doi.org/10.1007/s00170-016-8785-z

Yang, H., & Liu, X. (Eds.). (2012). *Software Reuse in the Emerging Cloud Computing Era*. IGI Global. https://doi.org/10.4018/978-1-4666-0897-9

Yang, S., Aravind Raghavendra, M. R., Kaminski, J., & Pepin, H. (2018). Opportunities for Industry 4.0 to support remanufacturing. *Applied Sciences*, *8*(7), 1177. https://doi.org/10.3390/app8071177

Zhang, C., Chen, Y., Chen, H., & Chong, D. (2021). Industry 4.0 and its implementation: A review. *Information Systems Frontiers*. https://doi.org/10.1007/s10796-021-10153-5

Index

Printed in the United States
by Baker & Taylor Publisher Services